Training Your Puppy Step-By-Step

A How-To Guide to Early and Positively Train Your Dog. Tips and Tricks and Effective Techniques for Different Kinds of Dogs

Paul Davis

© Copyright 2019 - All rights reserved.

The content contained within this book may not be reproduced, duplicated or transmitted without direct written permission from the author or the publisher.

Under no circumstances will any blame or legal responsibility be held against the publisher, or author, for any damages, reparation, or monetary loss due to the information contained within this book. Either directly or indirectly.

Legal Notice:

This book is copyright protected. This book is only for personal use. You cannot amend, distribute, sell, use, quote or paraphrase any part, or the content within this book, without the consent of the author or publisher.

Disclaimer Notice:

Please note the information contained within this document is for educational and entertainment purposes only. All effort has been executed to present accurate, up to date, and reliable, complete information. No warranties of any kind are declared or implied. Readers acknowledge that the author is not engaging in the rendering of legal, financial, medical or professional advice. The content within this book has been derived from various sources. Please consult a licensed professional before attempting any techniques outlined in this book.

By reading this document, the reader agrees that under no circumstances is the author responsible for any losses, direct or indirect, which are incurred as a result of the use of information contained within this document, including, but not limited to, — errors, omissions, or inaccuracies.

Table of Contents

Introduction 5

CHAPTER 1
CHOOSING THE RIGHT DOG 7

    PICKING WHERE YOUR DOG COMES FROM 8
    ENERGY LEVELS OF CERTAIN DOGS 12
    CHILD-FRIENDLY OR NOT 17
    INTERACTING WITH OTHER ANIMALS 19
    GROOMING NEEDS 24
    OTHER BREED SPECIFIC CONDITIONS TO REMEMBER 26

CHAPTER 2
PREPARING YOUR HOME FOR THEIR ARRIVAL 29

    STARTING POSITIVE HABITS FROM THE BEGINNING 29
    THE RIGHT HOUSE SETTING 32
    TOOLS NEEDED 34
    THE TRAINING ROOM FOR YOUR DOG 36
    HIDING CERTAIN BELONGINGS 41
    CHOOSING THE RIGHT TOYS 43

CHAPTER 3
PRINCIPLES TO TRAIN DOGS 46

    POTTY TRAINING 46
    BASIC TRAINING RULES 52
    POSITIVE REINFORCEMENT AND REMAINING PATIENT 54
    SPECIFIC RULES FOR SMALL DOGS 58
    SPECIFIC RULES FOR MEDIUM DOGS 59
    SPECIFIC RULES FOR LARGE DOGS 60

CHAPTER 4
TRAINING YOUR PUPPY OUTSIDE OF YOUR HOME 61

    GOING FOR WALKS AND LEASH TRAINING 61
    DOGGY DAYCARE 65
    PARKS 67
    FRIENDS' HOMES AND NEW ENVIRONMENTS 70
    WHAT TO DO WHEN YOU ARE NOT HOME 71
    PUBLIC TIPS AND REMINDERS 74

CHAPTER 5
SKILLS THAT DOGS NEED TO KNOW 75

    SITTING AND LAYING DOWN 75
    BARK CONTROL 79
    GREETING VISITORS AND WALKING PROPERLY 81
    KNOWING THEIR NAME AND COMING ON COMMAND 83
    FUN TRICKS FOR YOUR DOG 84

CHAPTER 6
KINDS OF EXERCISES FOR YOUR PUPPY 87

    TRAINING AND AGILITY CLASSES 87
    SWIMMING 89
    CATCH 91
    DOG-LED WALKS 92
    CREATING A ROUTINE 93
    INITIAL SHOTS 95
    SPAY AND NEUTERING 98
    PICKING THE RIGHT FOOD AND TREATS 100
    AVOIDING PEOPLE FOOD 102
    FLEA TREATMENTS AND BATHING 104

CHAPTER 8
WHAT NOT TO DO WHEN TRAINING DOGS 108

    AGGRESSION IN TRAINING 108
    INCONSISTENCIES 112
    USELESS REPETITION 113
    CONFUSING THE DOG 114
    ABANDONING A RITUAL OR HABIT 116
    TREATING ALL DOGS, THE SAME 117
    POSITIVE REINFORCEMENT WITH BAD BEHAVIOR 118
    NOT PROOFING TRICKS 119
    WAITING TOO LONG 120

CHAPTER 9
YOUR STEP-BY-STEP TRAINING PLAN 121

    GET TO KNOW YOUR DOG 121
    CREATE A SCHEDULE 121
    KEEP IT SIMPLE 124
    WORK ON REWARDS 124
    REPEAT, REPEAT, REPEAT 125
    CHECK-IN WITH YOUR DOG 125

CONCLUSION 126

REFERENCES 128

Introduction

One of the greatest joys in life can be the little pets that we have at home. Watching a dog grow from a teen puppy to a well-trained furry companion is an exciting thing. Many individuals believe that having a puppy is going to be a fun experience. It certainly is! At the same time, the hard parts of training are often overlooked. Some people can get in over their heads and end up with too much to handle when trying to properly train their four-legged friends.
You might be the most responsible pet owner out there. You could have only fresh organic treats and food, the top-quality grooming and veterinary care, and a personal trainer for your dog. No matter what you do, there are no circumstances that will keep you free from the slight chance that you might wake up in the morning to find a trail of diarrhea across the house, an expensive shoe that's been ripped up, bits of a dollar bill found in vomit, a school paper soaked in pee or a wad of toilet paper dug up from the trash can and ripped to bits strewn about a house.
Dogs are cute, but they are animals, which means they aren't free from having their nightmarishly disgusting moments. They might get fleas, ticks, and other creepy critters that can make your stomach turn. This isn't to turn you off from getting a dog! Nor are these aren't daily experiences. However, they are occurrences often overlooked by those wanting to adopt a pet, which is why millions of household pets end up in shelters every year. Don't think that having a dog is all disgusting, but don't think that you'll be free from these random instances.
Nothing we discuss is meant to scare you away or make you nervous about having a dog! With all the animals on the streets, in shelters or even in abusive and neglected homes, it is important to know everything that you are getting yourself into before adopting a pet. These are innocent animals who are going to look up to you from the moment you bring them home.

Don't think of your puppy as a human. If anything, think of it like a human baby, but throughout its entire life! They will be smart enough to learn new tricks, but they will also not realize that eating garbage is wrong. Throughout this book, we will give you a comprehensive guide of everything you need to know so you can successfully raise a healthy pup.

There are 8 steps to training your puppy:
1. Pick out your dog.

2. Prepare your home.

3. Start basic training.

4. Socialize them and prepare them for the world.

5. Teach them more challenging skills.

6. Ensure there is proper exercise.

7. Keep up with the health of your dog.

8. Avoid the mistakes and reflect on what you can improve.

Throughout each chapter, we will give you all the info necessary to help you create your training plan. In the final chapter, we will discuss how you can create a specific schedule along with other important reminders to live by throughout this process.

It is an exciting time and you can be assured once you bring your furry friend home, life will never be the same! Let's get started now in discussing how to choose the right dog for your home and your lifestyle.

Chapter 1

Choosing the Right Dog

Most people give up their animals because they can't afford them (Greenwood, 2016). At the same time, many individuals will end up giving their dogs away because they simply can't keep up with their basic needs.

Picking the right dog is important. There are a few options that you have to get started. You could go to a shelter, a breeder or a pet store. Once you have decided where to get your dog from, do your research on the specific breed. Throughout this chapter, we will give you tips on how to best choose the right dog, along with a few suggestions for people with specific needs. If you don't get a dog that matches your lifestyle, it can make training them more challenging, while weakening the relationship. Consider all factors that we discuss in this chapter when choosing the new family member you will be adding to your life.

Picking Where Your Dog Comes From

With over 900 million estimated dogs in the world, we can be certain there won't be a shortage anytime soon (World Atlas, n.d.). When choosing to bring a new dog into your home, you'll have several different options to choose from. There are a few important things to remember when picking the right place.

First, you want to ensure that you are getting a dog from a place you support. Before adopting a dog from anywhere, whether it be a puppy store, a breeder or a shelter, do some background research as to how they train their dog. Of course, the dogs aren't at fault for how they are treated, but it is best to make sure that we are supportive of those who care for their animals properly.

It is also important to remember that puppies will always be in higher demand, but do not overlook older dogs! We will be focusing on training your puppy throughout the book, but don't forget that contrary to the popular saying, old dogs can learn new tricks too! Even if you adopt a dog at 10 years old that isn't potty-trained, has a problem with barking, and has almost no structure, you can still train them. The difference between an older dog and a puppy in training is that puppies are simply easier to train.

When deciding on *where* you will be getting your dog, keep in mind that every year, over two million dogs are given to shelters. Not even half of them will end up adopted (ASPCA, n.d.). There are plenty of puppies in shelters as well. Just because a dog isn't the size of a toy animal doesn't mean it is not a puppy! A puppy is any dog less than a year old. Between their first and second years is when they are considered adults, but even then they won't always act like an adult.

After about six months and up until eight months, dogs grow out of their 'puppy' looks as well. They get bigger, their ears and face grow into an adult's, and their coats are usually fully grown in by this point. Though puppies are adorable, we will only have them as puppies for six months, so it is essential that we make the right decision for our future, as well as our dog's. You have several options from where you can choose a dog. These are the most common places that people get dogs, according to AnimalSheltering.org:

- Adopted from Shelter (44%)

- Bought from Breeder (25%)

- Pet store (4%) (Animal Sheltering, n.d.)

The remaining were smaller and random statistics, such as adopting from a friend, received as a gift from a relative or taken in as a stray.
Let's start by understanding what a shelter is. The first kind are municipal shelters. These are run by government officials and the animals are usually pick up off the street. Compassionate employees ensure population of certain dogs and cats are under control. They also ensure that sick and injured animals are given the treatment needed. These shelters are often overrun, so as a method of controlling population, animals will be euthanized after certain periods of time, or under special conditions in these shelters. Adopting a dog from a municipal shelter might be the best method to save a life!

There are private shelters as well that will take in animals under some of the same conditions. They will also adopt animals from caretakers that can no longer handle them, and they might even partner with municipal facilities. Private shelters might euthanize, but many strive to be no-kill as well. It is up to you to discover private shelters in your area and do your research to see what their adoption policies might be.

Humane societies and SPCAs will also have animals up for adoption if you have one in your area. These are more commonly known for their advocacy for animal rights, promoting education or aggressive neutering and spaying efforts to keep population down. Some are no-kill and others aren't. It is important to remember that even in no-kill shelters, overcrowding can occur, meaning that sick animals might not get the treatment they deserve.

Breeders are usually smaller family homes that make it their job to breed the dogs they have to sell the puppies in a new litter. Breeders will usually have one specific breed, giving you the option to seek out a type of dog specifically. Breeders are much more regulated than they used to be, and many veterinarians will have recommendations for you to find one which is more reputable. If adopting from a breeder, make it your mission to meet them and the puppy before adopting.

Sometimes you might have to travel far to do so, but it is worth the trip. Sometimes, puppies are kept far from any humans and live in shelters or sheds away from the home of the breeder. This is a sign that the animal might not have had enough socialization. Of course, we want to help all dogs, especially those in bad conditions. However, if you have a family of five children who are still young, an anti-social dog isn't going to be good for you. The best way to ensure that your new furry companion becomes a lifelong part of your family is to make sure it fits in with you and your life in the first place.

A final option you have is through a pet store. These are very controversial because of the use of puppy mills in order to populate the store. The Humane Society of the United States conducted investigations under cover and made disturbing discoveries while using hidden cameras. They reported that there were "significant violations, including sick and injured dogs who had not been treated by a vet, underweight dogs, puppies with their feet falling through the wire floors, puppies with severe eye deformities, piles of feces and food contaminated by mold and insects (The Humane Society of the United States, n.d.)."

The first thought is that these puppies should be purchased to help keep them from having to live in these conditions! Unfortunately, what happens is that this increases demand, meaning that more puppies would have to go through this. Puppies that aren't purchased from the store will end up being given away if they haven't been purchased after a long time. You are entirely in charge of your own decision to purchase from a pet store if you'd like, but as a responsible pet owner, you should be doing proper research first.

This is a big decision for you to make, which affects both your personal and family life. There are many benefits and disadvantages of certain places, and the welfare of dogs is highly sensitive in many societies. For this reason, certain places might be labeled as immoral, but at the end of the day, it will be up to you to decide!

Energy Levels of Certain Dogs

Now that you have a good understanding of where to get your dog from, let's talk about factors that you will need to consider when picking your breed! The most important part for picking your breed is understanding their energy levels. This is because it will directly affect the kind of care they need. Some breeds might need a huge yard that they can run in on a daily basis. If they don't get this time outside, it could increase their aggression or make it difficult to control them when they are excited. Other animals might only need a couple short walks a day and are perfectly fine in social settings. To make sure that you are picking the right animal, you'll want to consider how energetic they are.

Individuals who work long hours with jobs that take up a majority of their day-time should have lower energy dogs. If you can't take your dog for walks during the day and only have time for a quick potty break at night, a large, high-energy dog shouldn't be your first pick. If you have children who can help care for the dogs and who will be playing with the dogs frequently, a high-energy dog is perfectly fine. Single individuals who live alone would probably want a low to moderate energy dog. If you don't have access to a yard and live in a smaller apartment, a low energy dog is first choice. Having access to a dog park or a regular park and beach is great for your dog and can help provide them with a lot of freedom and energy.

Consider their energy levels for when you are training them as well. Some dogs will be able to train easily and will know right away what you are commanding them to do. Other dogs might be harder to control because they are so playful and active. Being active or not isn't something that is good or bad. It will all simply depend on your own personal energy level and how much you'll be able to contribute to the dog training process.

Lower energy dogs are those that are considered toy breed dogs, and the non-sporting group of dogs. Toy dogs have the lowest energy. However, this doesn't mean they aren't necessarily active. For example, a Pomeranian has low energy because you won't need a large yard for it to run, but it could still be the loudest barker in your apartment complex! Toy dogs were bred to be a companion to humans. They are specifically used to be carried around and sit on the laps of their owners. They will be best for those who are looking for a dog to have as a friend. These are best for older individuals, those with mental illness that might want an emotional support animal, or single individuals that aren't as active. If you are single or live with a partner or perhaps you live in an apartment complex, this is the best category for you. It doesn't mean that your dog just lays around all day either! All puppies are very active and will require training to prevent barking, jumping, and other high-energy characteristics. However, as time goes on, low-energy dogs will be the kind that become more relaxed and require less physical activity than medium and high energy dogs. These are breeds typically known for having low energy:
- Bichon Frise
- Boston Terrier

- Bulldog
- Basset Hound
- Bloodhound
- Mastiff
- Manchester Terrier
- Pekingese
- Papillion
- Irish Wolfhound
- Chow Chow
- Affenpinscher

High energy dogs are those that stem from the terrier, sporting, working, and herding groups. These are dogs that were bred to be by your side at all times. Hunting dogs were trained to retrieve the flying animals (duck, etc.) that their owners shot down. Herding dogs were trained to help round up farm animals such as cattle or sheep. Herding dogs love having wide ranges to run around, and they will certainly try to naturally herd humans, especially children.

Working dogs are those that have been trained for protection. They might also pull things like sleds or be trained for rescues. Terriers were frequently trained in order to help protect the home and hunt smaller animals. Dogs from these categories will want to swim, play games, and spend a lot of time outdoors. They are very friendly and intelligent as well. If you are someone who loves spending time outdoors, has a huge yard, wants a dog for a specific task (hunting, fishing, farming, etc.) then you will want a high energy dog.

If you want a dog with high energy, these are dogs you should consider:

- Australian Shepherd
- Shetland Sheepdog
- Russell Terrier
- Siberian Husky
- Dalmatian
- Border Collie

- Labrador
- Golden Retriever
- Beagle
- Poodle
- German Shepherd

Medium energy dogs are those from all kinds of breeds but are known for having a nice balance between being able to play outside with families, but also be fine with sitting on the couch with you while you binge watch TV. This is perfect for those who have a fairly active lifestyle but aren't quite the kind of person who spends the majority of their time outside. These are also the dogs that are perfect for families who want to include a pet but might not have as much time for their animals, like a single individual or a young couple might have. For medium-energy dogs, you might consider:

- Australian Terrier
- Alaskan Malamute
- Akita
- Bernese Mountain Dog
- Bergamasco
- Bloodhound
- Corgi
- Chihuahua
- Cavalier King Charles Spaniel
- Cocker Spaniel
- Dachshund

- French Bulldog
- Maltese
- Many Terrier breeds

Consider all aspects of your life and the way that you do things now. Getting an active dog doesn't mean that you will become active yourself necessarily. Dogs can do a great job at teaching responsibility to different owners, but it is best you get a dog that fits in with your lifestyle rather than trying to fit in with the lifestyle of the dog you choose.

Child-Friendly or Not

Having a pet for your family is a great addition. It is another member who you can share memories and joy with. It can teach you and your children responsibility, compassion, and kindness. Some of the best memories that children have include their pet! When considering a pet for your family, there are a few things that you will have to remember.

Consider the age of your children. If you have teens, then you can assume they will interact with a dog similarly to adults. Children between 10-14 might not be as active, so they can still be fine with a family dog that works for you as parents. If your child is younger or petite in size, no matter what their age, consider this if you are adopting a larger dog, as they might still end up being on the herding end of things!

For any family that has a young and active child, especially more than one, then you have to be very careful when picking out a dog. Dogs are sweet, kind, and loving. However, they also have the ability to attack and kill. Dogs don't need to be feared, but we have to remember that just like humans, they can act aggressively when provoked. Although you could train your dog and your child to interact peacefully with each other, when your back is turned, you never know how a dog might react if the child tugs on its tail, flails its arms, screams or does something else that could trigger an otherwise peaceful animal.

Dogs will act on instinct, and just like children, they don't always understand the right social cues. At the same time, remember that this is a fact for ALL dogs, so getting a child-friendly dog doesn't mean that you don't have to have a discussion with your kids about the proper way to care for it. Even a child staring at a dog for too long can be enough to give it mixed signals! These are the types of dogs that don't work well in households with children:

- Shih Tzu

- Siberian Husky

- Greyhound

- Pekingese

- Jack Russel Terrier
- Chihuahua
- Australian Shepherd
- Rottweiler
- Saint Bernard
- Akita

These dogs are less likely to have issues when you adopt them as puppies and raise them with your family. Again, this isn't to scare you away either. Most of these dogs are fine, and with proper care and training provided to both humans and dogs, issues can be avoided. These are simply dogs that are more likely to play rough, nip and bite, and growl or scare dogs. Many herding dogs will nip at the feet or ankles of those running around, so it is not always that your child is at risk for getting seriously injured, but they might learn to be fearful of the dog. Pick one that is right for the whole family and give you and your kids the chance to meet your dog and play with it before adopting.

Interacting with Other Animals

If you are adopting a dog, consider your other animals as well. We will discuss breeds that work best with different dogs but remember that some shelters might not let you adopt if you already have a certain number of animals. If you do plan to grow your furry family, ensure that all the pets you already have are up to date on their shots and you have documentation because there's a good chance you will be asked to provide proof of their medical history.

Cats and dogs have long been believed to be enemies. Part of this is because they are such different animals. Cats are perceived to be devious, sneaky, quiet, relaxed, and reserved. Dogs have always been perceived to be more active, excited, and friendly. One isn't necessarily better than the other, though this is a hot debate. They are entirely different animals with various needs and qualities that keep them as separate species. At the same time, having one or more of each in your home is a great addition to any family. Though they can be enemies, they can also be best friends.

The reason that dogs and cats might fight is because of their natural instincts. Cats are naturally more afraid of things that are bigger than them. They also need personal space and time to get to know new animals. If you walk into a room and pick up the cat sleeping on the couch immediately, it will not be so happy about this. If you walk into a room and start petting a dog, they're more likely to start wagging their tails and feel excited about your presence. Cats struggle with dogs because they aren't given that space!

Dogs struggle with cats because they are smaller and quicker, much like the rodents that they have been trained to attack. A cat might move across your house the same way an opossum or a raccoon might. Though they aren't rodents, they are still animals that move around just like the ones that dogs have been bred to capture.

Typically, it will be easier for you to bring home a dog and introduce it to a cat rather than the other way around. The cat might be angry at first, but after some hissing and hiding, maybe a swipe or two at the dog, it'll warm up. If you bring home a two-pound kitten and introduce it to your 80-pound Tibetan Mastiff, you might struggle to train the dog to be dominant whereas a cat would rather hide than try to fight new visitors.

Cats don't respect territory as much as dogs do either. Most cats are independent animals whereas dogs are used to being in packs. Of course, animals like lions are frequently found in families, but your household cat is very different from a pride of lions than it is from some dogs! Cats will walk up to you and sit on your lap whenever they want. Dogs are more territorial and when they see this, they might be likely to try and protect you from the cat. At the same time, you can have both in your home! The best kinds of dogs to bring home if you already have a cat are toy dogs and non-sporting dogs. This is because they will be more likely to listen to you, their pack leader, and trust that this animal isn't a threat. Obedient dogs that are well-trained will typically do fine with other animals anyway, but still consider your cat before bringing home another pet. Listed below are the dogs that will do the best with other animals, and even cats:

- Maltese
- Boxer
- Bichon Frise
- Cocker Spaniel
- Basset Hound
- Labrador and Golden Retrievers

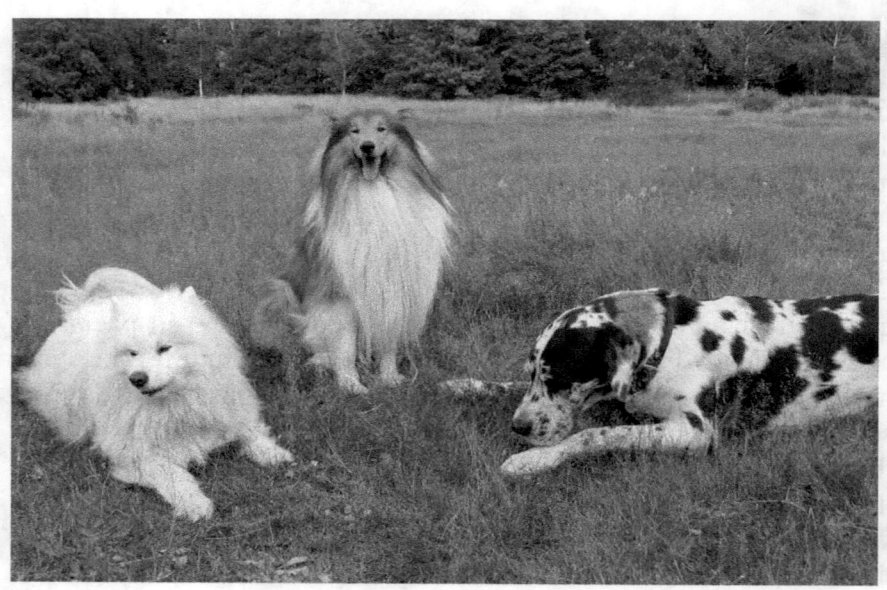

If you are considering having multiple dogs, you will want to consider different breeds that work well together, and those that don't. Many households will simply get dogs that are from the same breed, and this is the safe choice. We should still consider those breeds that don't quite mingle the best together.

First consider your dog's past and if they have gone through anything negative with another dog. For example, let's say that you already have a mini poodle at home, and you want to introduce another dog. There is a pug that lives in the house next door that you think is cute and you'd consider buying. However, the pug barks and growls at your mini poodle every time you go outside together. Though the pug might not have anything against either of you personally and is just being territorial, your mini poodle will still have a negative perspective about pugs. They will associate their ear shape, face shape, and body shape with something scary, so don't bring anything home that looks like a pug in this case.

Some dogs are more territorial and are the 'alpha' dogs. If you have a strong territorial male dog already, do not bring home another male dog. You would instead consider introducing a female since they are less likely to be aggressive. For multiple dog homes, you will want to consider males more often as singular pets or only have one male with other females, and have multiple dogs be female. This isn't the case every time. there are plenty of households with multiple male dogs that do just fine. In any case, if you are picking out new dogs to introduce to the ones you already have, don't overlook this point.

Small dogs will tend to be a little more aggressive toward big dogs. They will feel threatened and as such will show their dominance to the big dogs since they feel as though their size might be lacking. If you have a Pomeranian at home and want to bring home a Chow Chow, then you just might end up finding that the Pom doesn't stop barking and continues to try and exert its 'alpha' stance.

Consider dogs with matching needs over all else. You don't want to get a Pekingese, who could sleep up to 2o hours a day, and then bring home a Labrador retriever who will need hours of play. If your lifestyle is this diverse, it could work, but you want your pets to be on similar schedules, so it is best to consider that their needs will match each other.

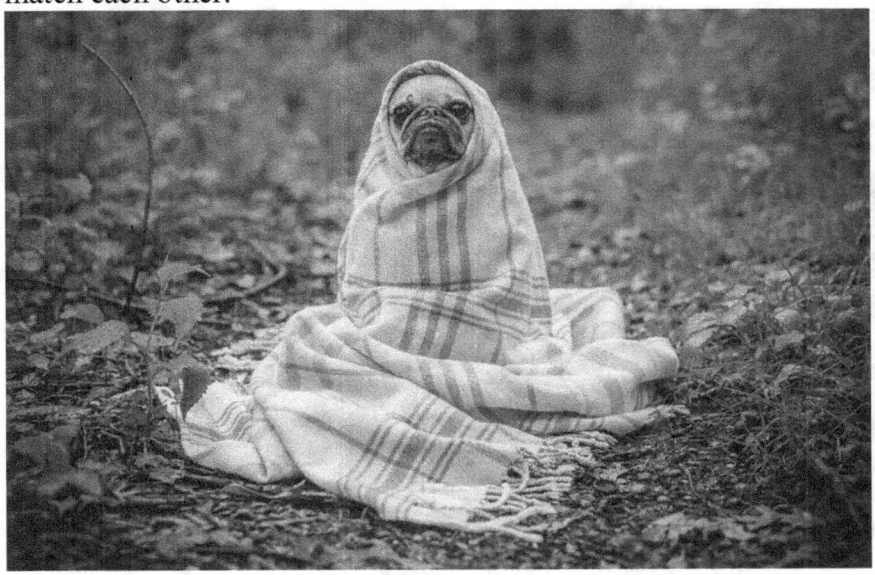

If you only plan on having one pet, and one dog, that is perfectly fine! There are many dogs that just want to hang out with you and you only. The dogs that prefer to be alone the most are:

- Great Danes

- Collies and Sheepdogs

- Poodle

- Pug

- Chihuahua

Grooming Needs

Some dogs are fine with a quick rinse and a brush, but others need bi-weekly or monthly grooming. If you get a dog that needs frequent grooming, but you don't have any time to groom them, they can get knots, matted hair, and start getting stinky much faster. Of course, many individuals like the look of groomed dogs and wish to have one for aesthetic purposes. Grooming can be expensive and tedious, so you'll want to consider the different grooming needs of various breeds.

The dogs that need the most grooming can also be the ones that are best for certain families, however. Hypoallergenic dogs are those that don't shed as much, so they are best for people with allergies. At the same time, since they don't shed, their hair will get rather long. If you don't properly groom them, their fur can get knotted. Ungroomed hair can collect sweat, dirt, and bugs. This could harm your dog's fur and make it easier to spread pests inside your home. If a dog has a knot, they can still be bothered by it, so they might chew and scratch, causing a more serious injury than the initial knot.

On the other hand, if you have dogs that don't need regular grooming, at least from a professional, they will still need some form of care and maintenance. All dogs will still need an occasional bath, and it is important to keep up with their nail and tooth hygiene. We will cover that in a later chapter. Dogs that are non-hypoallergenic shed their fur rapidly because they are constantly growing new hair. This means you will have dog hair on your clothes, furniture, and everything in between.

These are the dogs which will require the most grooming:
- Poodles – all variations (mini poodles, cockapoos, etc.)
- Bichon Frise
- Shih Tzu
- Maltese
- Yorkshire Terrier
- Mini Schnauzer
- Lagotto Romagnolo
- Irish Water Spaniel
- Cairn Terrier
- West Highland Terrier
- Scottish Terrier

These are the dogs which won't require frequent haircuts, but will shed a lot:
- Corgi
- German Shepherd
- Australian Shepherd
- Labrador Retriever

- Siberian Husky

These are dog breeds which will frequently shed and also require grooming and haircuts. These are not recommended for those with allergies or those who don't want to keep up with grooming:
- Akita

- Chow Chow

- Collie

- Pomeranian

- Pekingese

If you want to get a dog that requires grooming, of course, you don't have to know how to do this yourself. It is quick and easy to drop your dog off at the groomer's or pet shops that provide grooming services. This can actually help socialize your dog when done frequently. The biggest downfall of this is that you need to find the right groomer, which can sometimes be expensive. The more work your dog requires, the pricier it will be. If you are already struggling financially, don't get a dog that will need regular haircuts.

Other Breed Specific Conditions to Remember

Many of us have an idea of a breed that we want years before we even get a dog! All dogs are special, and they each have their own personality traits you'll want to consider. There are a few other factors you have to remember when picking a dog. Some people will give dogs to shelters over simple things, so to ensure you keep your dog around for the long term, remember some of these important aspects when deciding a dog to pick:
- Breed-specific illnesses
- Dogs with separation anxiety
- Dogs more prone to health issues

Of course, we want our dogs to be healthy. Because of years of breeding, there are certain health conditions that some dogs have gained throughout the years. While it is incredible what we have been able to do for dogs including the way that breeding helps them have certain characteristics, it is also essential to consider how breeding has led to more serious health issues.

Larger dogs, such as German Shepherds, might be more prone to hip dysplasia. Boxers are also at risk for elbow dysplasia, among certain cancers and heart conditions. The same can be said for Labrador Retriever. Both Labs and Golden Retrievers are also more prone to be overweight, which could put them at an even higher risk for health-related issues. These dogs require a lot of exercise, but it is also important that we don't put too much pressure or strain on their bodies to keep their joints intact and healthy.

In addition to some of these health conditions, larger dogs are also known for not living as long as smaller breeds. Part of this might be because of the way that they have been bred. They grow so rapidly that it ages them faster in the process. The larger the dog breed, the less likely it is to have a longer lifespan. This is sad in itself but remember the impact it could have on your family. Some large dog breeds, such as Great Danes, might only live to be five to eight years old (Soniak, 2013).

Smaller dogs are known for living longer since they age much slower. They might live up to be 15 years old depending on the breed. However, this doesn't mean they aren't still prone to disease and illness. Some smaller dogs, like Yorkshire Terriers or Miniature Poodles are more prone to have issues with their gastrointestinal systems. Part of this might actually be because of the stress they experience! Some smaller dogs are more likely to have ulcerative colitis, which is when the intestine or colon, becomes inflamed. It is more treatable, but when untreated or unmanaged, it can cause more serious health issues.

Certain breeds, like Pugs and Bull Terriers, have been so bred throughout history that they are very likely to experience certain health issues. They might have trouble breathing, could become dehydrated or exhausted faster, and will even snore at night in some households (Kiefer, 2017)!

All these dogs are equally cute, special, and important. It is still crucial that we consider their medical needs, because vet bills can be massively expensive. If you can't afford to pay for certain surgeries upfront, you could stand to lose your pet. Some veterinarians will work with you on creating payment plans, but many will require that you have several thousand needed for surgery upfront. We should do everything to care for our pets, but if you are not in a position to pay $500 or more in vet bills unexpectedly, it is important to reconsider getting a pet.

Aside from issues with health, some pets will be a little needier than others and require more attention and care. Some have separation anxiety that could cause them to have deeper issues. No dogs enjoy being home alone, but they will get accustomed to it in due time and with proper training. However, sometimes, being left alone for hours at a time can cause them to have some serious anxiety which can mean they may have accidents indoors, chew at furniture or objects, and whine or scratch at the door for hours on end. Not only is this annoying for neighbors, but some dogs have actually scratched at doors to the point that their paws will bleed. Consider your lifestyle and ensure that you don't get a dog that doesn't do well snoozing on the couch alone for a few hours while you are at work. If you do get one of these dogs, make sure you have it in your budget to hire a dog walker or sitter to check in on the dog during the day while you are at work. Listed below are the neediest dog breeds that do not like to be alone:

- Cavalier King Charles Spaniel
- Australian Shepherd
- Labrador Retriever
- Jack Russell Terrier
- Border Collie
- Miniature Poodle
- Bichon Frise

Chapter 2

Preparing Your Home for Their Arrival

Your dog will no doubt be scared when you first bring them home. They're coming into a new environment they've never been in before, which can be rather frightening! Set them up for success by using care and consideration when setting up your home. Whether your dog is three weeks or three years when you introduce them to their new space, there are a few things to consider. Not only is the environment in which you'll be training them important, but the tools that you have are special to consider as well. Stay positive, create healthy habits, and hide anything you don't want in the mouth of your pup! Let's dive deeper into these topics so you can be as prepared as possible.

Starting Positive Habits from the Beginning

From the moment your dog walks into their new home for the first time, you should do your best to train them to have good habits. If you don't start right away, then the dog won't be able to pick up on these healthy habits you are trying to instill so you can elicit good behavior. Dogs are very ritualistic animals and prefer to have a routine. Not only is it their way of operating within our society, but it also gives them peace of mind. Dogs can be very anxious creatures, especially if they can't predict what will happen next. A routine gives them the chance to understand what they might need to be prepared for and the ways that they can best go along with the natural flow of things.

Don't put off any vet appointments and necessary shots. Your puppy might seem perfectly healthy with no issues whatsoever. You might not be an active person or someone who would take your dog out in public anyway. In this scenario, it can be easy to think, "Nothing bad is going to happen, the puppy will be fine." Chances are, this is true. However, after all the love you give to your dog, this should be an extra step to ensure that they will be around for a long time. You never know what other dog they might interact with who could spread something. They might get one flea bite which could have been from an infected insect. There's no point in risking your dog's health, so keep up with necessary vaccines from the moment you bring them home.

Make sure to purchase all that is necessary before getting started. The dog should be the last addition that you have for your home and family. They should have all the necessary toys, bowls, leashes, collars, harnesses, beds, and everything in between to ensure that they will feel comfortable in their new home.

Practice what it will be like to have a dog before you even bring your four-legged friend home. It can seem tedious, but it is a good way to ensure that you are really prepared for the routine. Start by making sure that you keep all your belongings picked up. Puppies are like toddlers and can be great at finding random objects to chew on or play with. Every three to four hours, make it a point to go outside for a quick walk. Make sure that you get in the habit of establishing set times and routines so you can leave the house for about 10-15 minutes every three to four hours, just as you would once you start walking your dog on a regular basis.

Check-in with your emotions, because dogs require a lot more than just physical energy. Make sure that you are a patient and compassionate person. If you are someone who struggles to keep their cool, especially in stressful situations, it will be good to come up with some coping mechanisms because puppies will certainly test your patience. There will be moments where they have accidents, they won't always fully understand what you are trying to say to them, and they'll cause frustration among some family members. This isn't a bad thing, but it can be challenging to deal with, especially for individuals who don't have their emotions in check.

Be prepared to put in the work all day long. Even though your puppy will have plenty of moments of playing, sleeping, and just relaxing, you will still have to be alert. If you can't see your puppy, you'll have to find them to ensure they didn't get lost under a piece of furniture or in a different room. You'll have to be prepared to wake up early to take them out, fill their water bowl, and feed them. If any of this sounds challenging, then there needs to be more preparation done on the owner's end before bringing a new puppy into the home.

Make sure that you have the proper discussions with roommates and other household members. Decide who will be in charge. For example, if you have a roommate who you aren't very close with and who isn't really interested in dogs in general, you can't expect them to be a caretaker. If you are bringing a dog home to a family with a husband, wife, and three kids, then you'll probably want to split the responsibility equally among everyone. It is important to have a talk about all responsibilities beforehand. You can create a schedule for family members. Maybe each day, a new person is responsible to take the dog outside. Perhaps one person is in charge of feeding the dog and everyone else has to take turns walking the dog. You have to have these discussions before, so you are prepared instead of being stressed and confused in the middle of it all.

There has to be consistency when training the dog. That is key. You'll have to make sure that there are house rules. Mom cannot let the dog on the couch if dad wants it to always sit on the floor. The kids cannot give the dog any snacks that they're eating if the rule is that the dog only eats dog food. One family member cannot have a rule that the rest of the family members don't follow because this will hurt the dog in the end.

Traveling isn't going to be easy anymore. Make sure that you don't have trips coming up, and if you do, ensure that there's someone there who can take care of the pup while you are gone. Make sure that you have a schedule so that the puppy is never alone for more than a maximum of three hours. For puppies younger than 12 weeks, it should really be no more than an hour. Once they are older than a year, and in some cases six months, they can stay home for longer and up to six to eight hours depending on the dog breed.

There will be a lot of changes. Nothing will ever be the same. Some things you'll be prepared with, others will certainly surprise you. Don't underestimate just how prepared you have to be!

The Right House Setting

Your home is your puppy's home. It has to be a welcome and friendly environment in which they can thrive in. Imagine that you are the puppy before bringing it into your house. You are currently with all your brothers and sisters somewhere trying to figure out how to walk, run, and even bark. Then, some new people you never met pick you up and carry you away. They bring you to a place that you've never been to before. What would you want in this new place to make sure that you feel right at home? We can't project emotions on puppies like humans, and animals don't stick with their families all their lives anyway. What's most important to remember is that they will be frightened because things are suddenly different from how they just were. Ensure that your puppy is given all it needs to feel right at home from the moment you bring them home.

The puppy needs its own space. As it grows, you won't have to dedicate an entire room to the dog. You might be able to eventually just have them roam the house, even when you are not there to watch over them. As a puppy, you should set aside a dedicated area where they can run freely, sleep, and nap during the day as you are with them. Don't give them free range of the house just yet. That will be too overwhelming, and they'll end up getting themselves into a place that you didn't even think of at first. Your puppy will want something familiar, so use baby gates to block off a section of the house that they can feel the safest in.

Give your puppy a place to play that doesn't have a rug that you don't want ruined. You can't bring a puppy home and into your freshly carpeted living room, only to get mad that it immediately pees on the carpet. If you do have a carpeted room, invest in some thrift store sheets or extra rugs to put over the floor to protect it from the puppy. At first, your puppy won't defecate or urinate in massive quantities that are hard to clean. It is still important to be prepared for them to go 'potty' wherever and whenever they have to as you are starting to potty train them.

You'll want something steady and constant for it to bond with. Create a little room in an office, a bathroom or laundry room for your pup. These are smaller areas that they can go in for naps and get away from it all. Block off stairs just like you would for a newborn baby. Anywhere that a small creature could get hurt is an area that needs to be entirely inaccessible to your puppy.

Use puppy pads to create a path to the back door or the door that they'll be regularly going out to go to the bathroom. Let's say you are creating a space for your puppy in the sunroom that is adjacent to your living room since there is a direct path toward the backdoor on the other end of the house from your living room. Additionally, you spend most of your time hanging out in the living room, which will make your puppy feel less lonely when they spend time in the sunroom. You put a little bed in there, their crate, and some water and toys. Use newspaper or purchased puppy pads to create a path all the way from the sunroom to the backdoor. Each time they have to go out, let them walk along this path so they get a feel for where they're supposed to be going to the bathroom.

Make sure other animals have their own spaces during this time as well. Your cats will especially be more scared, so you will want to put their litter box or food bowls in a private space so that they can have alone time as well. Don't do anything that will freak them out, but if their litter box is in the laundry room where you also plan to keep the puppy, then they might not want to go in there, and could end up peeing or pooping in a random spot in your home. Consider all the animals and their need for their own room, not just your puppy.

Tools Needed

Once your home and your family are prepared to welcome your new furry companion, it is time to start stocking up on the right supplies. The best place to purchase supplies are from a certified pet store. You can shop online as well. When you go to certain big stores that stock cheap and generic dog supplies, you might discover that the products aren't as high-quality. This is fine if that's in your budget, but there should certainly be an emphasis on purchasing the highest quality items for your dogs. It may seem expensive initially, but if you really think about it, you will only need to buy most of these supplies once. For example, you'll purchase food and water bowls for them, but it is not like you'll be buying a new bowl every month. Every year or so, it is good to replace things just because you never know what bacteria or other germs they're holding onto that don't get washed enough, no matter how much you clean and disinfect them. However, most of these things will be a one- or two-time deal to buy, so now is the time to splurge and spend your money on your four-legged friend!

The first thing you'll want is a collar, harness, and leash. At first, puppies won't be able to run far away from you, but this doesn't mean you shouldn't still train them on a leash early. Make sure that you buy a collar specific to their size. You should be able to fit two fingers sideways into the collar to make sure they have enough room to breathe and grow. Get them the appropriate tags as well. Their name, along with a phone number and even your name is a good choice so that others can call them and call you if the puppy ever gets separated and lost. Tags with their vaccinations will also be helpful so that others know if the puppy is at risk of carrying or catching any diseases or illnesses.

A harness is a good choice as well. Puppies especially will tug and pull on their leash, and only having a collar might mean that they hurt their throats frequently. As you start to go for walks, the puppy might frequently get into things they shouldn't. Maybe there's an old sandwich or chicken bone that was tossed on the street from someone who was walking by earlier. This is a big find for your puppy and they'll try to devour the snack! Rather than jerking their necks back on a collar to stop them, you'll have a harness that could even help you to pick them up quickly.

Training pads will be a good addition as well. These are items that don't necessarily need to be the most expensive brand, and many stores carry generic versions that do just as well. The point is that they absorb moisture and provide your puppy with a scent to let them know that this is the place they should be using the bathroom, nowhere else. These are not to be used in replacement of going outside. If you want to train your puppy to use pads even into adulthood as they become full-grown dogs, that's perfectly fine. Some people might live on the 20th floor of a building and going outside three to four times a day isn't an option, especially late at night in the big city. They might want to use pads, and that's fine, it is totally up to the individual. However, if you do plan on training your dog to only go outside, the pads should be for accidents, not the only way that you teach them to go to the bathroom. They should be used during a transition period as the dog becomes accustomed to going outside.

You need to also purchase dog food and treats, but we'll get into the specifics of this later. As far as food bowls go, stainless steel is always the best option. Plastic bowls can trap moisture and cause bacteria to grow, potentially making your dog sick or even giving them little pimples on their chin! If you are using a glass or porcelain bowl, it could end up breaking if your puppy gets too rough. Stainless steel is best, especially since it can be cleaned very easily. Purchase one that has a rubber bottom to keep it from sliding all over the place as your dog eats. Keep it low at first for the puppy, but depending on the breed, a raised food bowl might be better so that they don't have to strain their neck or back to reach down to eat.

Have some form of a pet bed for them. It doesn't have to be a big dog bed, it could simply be a cardboard box with some blankets and a pillow. You could give them a few different options as well to see which bed they like the most. They simply need a place where they can sleep all alone and do so whenever they feel like it.

Toys are a big requirement! You want to teach them to play with certain objects and not to play with things that are more important to you. We'll help you discover the way that you can find what's best for your furry friend. Let's discuss now the option of using crates as a training room for your dog.

The Training Room for Your Dog

You have two options for your dog's 'room.' You can use a crate or you could have an entire room dedicated to them. A crate is preferable to most families, but if you can dedicate an entire room such as a spare bedroom, this is always ideal. Don't use something like a mudroom where people will be passing through frequently. For this section, we are mostly going to be discussing a crate as your dog's training room.

The reason that your dog needs a room is because they are den animals. This means that they are used to finding small holes and caves that they can associate with being their own. Think of animals in the wild that are like dogs. Wolves frequently find caves which they can easily hide in. They might find an old animal nest and use this as well! Your dog's room can be like their crate. If you choose to use a crate, then you will want something much bigger than the dog. For example, look at the photo above. It looks like it could probably fit into an even smaller crate, but it shouldn't. The dog should be able to stand up and fully turn around in their crate. If they can't, it is beneficial to purchase something a little bigger for them.

This room is not something that you should use to simply shut them away in. It has a specific purpose. This is the room your puppy will use to relax, play, eat, and sleep in, along with their puppy training. It is not a form of punishment and it should never be something that you use to isolate them. It is a safe haven, a place they want to go, and somewhere they feel completely comfortable in.

This is the room that they will go to when they need to sleep or when you aren't going to be home. A place of comfort and familiarity. If they have been playing for over an hour and you trained them throughout, you can put them in the crate as a way to let them know that this is a quiet time.

Never use their crate as punishment. If they go inside the house, chew on something they're not supposed to or engage in any other negative behavior you want to punish, don't put them in their crate. They won't be able to understand that what they did is wrong so therefore they need time to reflect and think about things. They will simply see that you are upset and confused about the situation.

Don't put your dog away for more than three hours at a time when they are younger than six months old. As your dog becomes an adult with more bladder control, then you can crate for up to eight hours, but this should really be the maximum. This should only be in rare circumstances when you can't find someone to help them out, not the entire day. If you are someone will be gone for eight hours and beyond, using a crate shouldn't be your puppy's room. You can use the crate and give them the option to go in, but a bigger space is better if you ever have to leave them for that long at a time.

Your dog wouldn't go to the bathroom in their den, so don't expect that they'll just go if they really have to. Instead, they will hold their bowel movement to the point that they physically can't anymore. The den isn't a place to put them to go to the bathroom. It is an area where you can train them to sleep, eat, and relax. If you want to leave them home all day and don't care if they roam freely in your home, then using an entire room with puppy pads is a better option to keep them away.

This is not prison. Only keep your dog in its crate as you would want to go in your room. If you make them fearful of the crate even once, they will be scared of the crate every time that they go in it afterward. This is a place where they can be alone, have their private time, and stay completely relaxed.

Even if you want to sleep with your puppy, they should go through a crate training period. This is to help them control their bladder. At night, you can put them in the crate and be sure that they will try to hold it in all night. If you let them sleep with you, they might jump off the bed and wander to a spot to relieve themselves, then come back to bed without you ever knowing (until you find the mess, of course)! After a few weeks of crate training at night, you can start letting them out of the crate and see if they are able to sleep with you while holding in their bladder movements all night long.

At first, this room should be somewhere they can easily access where you and the rest of your family will be. Don't put their crate in the basement where they can't get to it. Keep their crate in the living room where everyone hangs out or the kitchen where you might always be cooking. This is because you don't want to create the idea that their crate means isolation or that they can't be around others. If you get them used to them being in their crate while you are still there, it will help them better understand that this is their home, not their punishment. Now that you understand the importance and meaning of this small space, let's discuss how to introduce them to the crate. Again, if you don't want to use a crate, a small room is just as fine. We'll be discussing how to use the crate and introduce this idea to your puppy, but by all means, apply this to a smaller room as well.

You'll start by introducing this crate to them just as you do everything else – slowly and led by them. Put some toys in there and let them go in on their own. Maybe you place a towel in the crate that has the texture of their mother to make them feel more comfortable. Some people have even tried to put clocks in there to mimic a heartbeat. You might even put a treat or two in the crate. Let them go in on their own and explore this place for themselves. It will be a place they can get to know and explore, not just an area where they are shut away.

Never force your dog in the crate if they are actively resisting. If it seems like it is something scary, be patient. Sit next to the crate and play with toys. If they aren't into the crate, don't pick them up and put them in. If they go in just with one paw at a time, don't push their butts trying to get them to go faster. It might take a few days for them to actually go in, but that's totally normal. It will be better for you and your puppy if you are patient in this process. If they don't go in, place a toy at the front of the gate, not all the way at the back. Let them get this toy out if they want, and even if they don't go in the crate at all, it is still helpful to make sure that they are slowly transitioning into this.

You can make this transition easier by giving them meals in their crate. Start by introducing them to the crate on the first day. On the second or third day, once they display more curiosity, put their food bowl right inside the crate, just at the front of the door so they could even eat their food without going all the way in. Each time you give them a meal, push it a little further from the door so they have to go deeper into the crate. Eventually, you can put the food all the way to the back of the crate.

This is when you can start to close the door while they are in the crate. Don't slam the door and startle them. Slowly close the door behind them and see how they react. If they panic, open the door immediately and use positive reinforcement once they come out. If they don't even notice, congratulations!

Keep your dog in for 10 minute intervals, but nothing longer than that. After they've finished eating, let them stay in there for a few minutes, then let them out. Do this during every meal time until they don't show any signs of being afraid of their crate. Start leading them in without meal times after this. For the first week, do so every day. Practice letting your dog in the crate and keeping them in there for at least five minutes, but not much more than 10 minutes. This is to get them used to it. You should always be there in the same room during this training period. You don't have to be engaging with your dog. In fact, you should do the opposite. Watch TV, read something or just relax on the couch by them. Let them see you, but don't feel the need to look at them and play with them while they're in the crate. After these few minutes, let them out and use encouraging reinforcement the whole time, so they know that this crate is a positive experience.

Don't let your dog out because they are whining. Wait until a few minutes have passed after they've stopped whining to let them out. If you don't wait, they will associate whining as their key to getting out of the crate! It is hard to watch your puppy cry, but it is okay! It is better for them in the long run.

Practice this crating method before you actually leave them for long periods. Until they're used to a crate, you'll want to be home with them. This is why you should take a few days off work to make sure you are home with the new puppy.

You should be crating them when you leave and when it is nighttime, but not only these times. Crate them when you need to clean the house, or if you need to get some work done. Shorter periods are better, but if you only put them in the crate when they're alone, then it will make them associate the crate with isolation (The Humane Society, n.d.).

Hiding Certain Belongings

Puppies are like babies. They will explore and try to put things in their mouth as often as possible. It doesn't matter if it is a $300 pair of shoes or some poop from the litter box. If it has a scent, the puppy will try to see if this is food! You have to watch your puppy like you would babysit a child roaming around your house. A puppy has heightened senses, so they'll be even better at picking things out that shouldn't be eaten!

Make sure all electronics are picked up. This includes wires connected to the TV. Any cords or cables freely around the house should be put up. Use plastic hooks that can easily be removed from your walls to help keep wires and cords off the ground. Not only is this so they don't chew on them, but you never know what area of the floor might have a scent that triggers them to relieve themselves right there. If you can't put the cords up for whatever reason, you can use tape to keep them stuck to the floor and protected from your pup.

Block off electrical outlets. If you can't use a specific plastic child/pet safe outlet cover to protect them, use a baby gate to create a barrier to heavy electronic areas in your home.

Ensure that anything you don't want in your dog's mouth is put away. This is anything from shoes to throw pillows and chair legs, and so on. If you don't mind a little puppy slobber on your furniture, then keep it around, but anything else, make sure to hide from the pup. This won't be the case the whole time. Once your dog learns some rules and boundaries it'll be easier to ensure that they aren't getting making a mess or chewing on what they shouldn't be.

Check for loose buttons on furniture, thread getting pulled out, and other things that might be easy for your dog to put in its mouth.

Consider the cleaning products you might have used on the surface of certain things. Did you just bleach the bathroom? Did you use wood polish on the furniture? Your puppy might try to chew on these surfaces, and you don't want them accidentally licking up any harmful chemicals.

Plants can be an area of interest for pets. Ensure that any plants within the reach of your puppy have been put away, or that they are at least safe for animals to chew on and not something specifically poisonous to dogs.

Keep other toys that are meant for bigger dogs away from your puppies. This might be bones, big rubber items, and anything else that can be hard on their teeth or cause them to play too rough. Puppies need puppy toys! Consider things that you would never assume your dog has interest in that they still might. You might think, "That's fine sitting there, why would my puppy have any interest in that?" only to discover that it is in their mouth five minutes later.

Of course, there are many toxic foods that your dog should stay away from. These include:

- Chocolate

- Caffeine

- Grapes

- Onions and garlic

- Alcohol

- Raw meat

- Meat bones

Dogs love to chew on bones, but these should be safe bones purchased specifically for dogs. Chicken bones are more brittle, especially once cooked, so they could snap in half and your dog might end up choking on it. Even if you think you have managed to pick everything up you need to, you might discover that the puppy helps you see things that weren't there at first.

Choosing the Right Toys

All dogs deserve to have plenty of fun toys to play with! Toys help them learn how to interact with others while also providing them some entertainment. Dogs need to feel like they have a purpose. They like tasks because it makes them valuable members of their pack. They prefer to be outside chasing mice, birds, and other little critters to bring to their pack, but that's not ideal for us as human owners! Provide your puppy with all the toys they need for the best development possible. The best option for purchasing toys is to do so at a pet store. Be mindful of the quality of toys you purchase such as durability and materials.. Cheaper toys might have plastic or fragile rubber that breaks apart after just a few minutes of chewing.
Don't assume kids toys are safe for dogs. Just because a baby can play with it doesn't mean that it will be safe for your dog too!

Use toys that are specifically toys. Don't give them an old pair of shoes just because you don't care about the shoes anymore. They will think that all shoes are things that they can play with, not just the specific pair that you gave them.

Consider toys based on what your dog's breed is. Some dogs are hunting dogs that care about catching game. Toys that mimic animals like birds and ducks could be good for them. Other dogs are more interested in running and chasing, so they might do better with balls and toys that you throw long distances. You can never go wrong with a game of fetch! Many dogs like the same kinds of things, but it is also important to check with your specific breed and understand what their favorite things to do are.

Puppies want soft things they can roll around with. Their teeth are still developing, and they'll also want things that mimic their brothers and sisters. Puppies play with their siblings to learn how to function as adult dogs. They need that interaction to help them learn and grow.

Before they're six months, they will be teething pretty hard. Rubber toys and other things they can chew on will be best for them. Older dogs should have softer toys as well, as their teeth will be more sensitive.

For large dogs, don't give them anything may break right away. Even though it might be a specific plush toy made for dogs, your pup might have tough enough teeth to rip through anything they want.

Give them a variety of toys, not just one. This will help you discover what they like and dislike as well.

Dogs can have puzzle games too! You might get them a ball that has treats in it, forcing them to move the ball around to have treats fall out.

It is important to have toys that they can play with, however you want to ensure they socialize with kids as well and can play with them too. Some ways to ensure this are by including ropes that you could play tug-o-war with and having your kids throw some balls that they can fetch. Now that you have all the supplies you'll need to help your dog grow and develop, let's look at the actual training practices you will be teaching your dog.

Chapter 3

Principles to Train Dogs

You have your puppy, you have all purchased all your supplies to ensure your puppy is happy and comfortable, and have gone through any other measures to ensure they are healthy and safe. You are prepared. Let the training begin! There are a few essential principles that will be important to teach your dog. You should follow these rules and do these things as best as possible to give your dog a chance to be as happy and healthy as can be. Potty training will be the first thing you teach them. This is because it is a basic need that they have, and there's no delaying their natural urges. In this chapter, aside from potty training, we will give you a few basic training rules as well as methods of positive reinforcement to use throughout the training process.

Potty Training

This is the most essential thing that your dog will learn. We will spend a lot of time in this section, because this is one of the biggest headaches for new dog parents. A dog that chews on some furniture isn't a big deal, especially because puppies haven't developed strong teeth yet. Having to teach your puppy some boundaries can be simple as well. The biggest headache is when a dog goes to the bathroom inside the house. It is stinky, inconvenient, and generally annoying. Don't worry or fear this though! It is still natural, and plenty of dogs are easily trained to the point that they don't urinate or defecate inside at all and haven't for years at a time.

There are a few important things NOT to do to make this transition much easier you are your pup. First and foremost, never push their face in pee. This idea had been around for a long time as a way to train your dog. People thought that if you made your dog uncomfortable in this way they wouldn't go to the bathroom inside the house again. This is completely false and could even delay proper training. You are simply teaching your dog that you are dominant and that they should be ashamed for going to the bathroom.

Making sure that your dog is properly trained isn't just about taking them out quickly when they have to go. To help prepare them and set them up for success with potty training, ensure that they are on a consistent schedule and your communication is consistent as well. Dogs need regulation. They don't know the difference between inside and outside. They know the difference between going to the bathroom in an area that's acceptable and somewhere they shouldn't go. The only way that they'll know what's right and wrong is by you teaching them the boundaries they have to follow.

You'll want to ensure that you help them have a schedule by feeding them at the same time every day. If you are feeding them at the same time every day, this also means they'll likely be going to the bathroom at the same time every day. It will be easier to predict their schedule so that you know when they need to be taken out. The dog will appreciate this as well because it will make it easier for them to hold their bladders when they have to.

Having a crate is the best option in the process. They don't want to soil in their cage. They like to keep their environment clean just as humans do. Cleanliness isn't about appearance. Most animals are aware that it is simply not good for your health to sit in urine or feces. Never let them go in the crate or expect that they will go in their crate just because you leave them in there for a long time.

Your puppy will be able to hold it for as many hours as they are in months, until they reach nine months old. This means that:

- A two-month old puppy can hold it for two hours.

- A three-month-old puppy can hold it for three hours.

- A four-month-old puppy can hold it for four hours.

- A five-month-old puppy can hold it for five hours.

- A six-month-old puppy can hold it for six hours.

- A seven-month-old puppy can hold it for seven hours.

- An eight-month-old puppy can hold it for eight hours.

Eight hours is the maximum that you would want your puppy to be holding it for. Anything longer could hurt them. They might be able to hold it for longer, but that doesn't mean that it is healthy for their body overall. Some puppies do take longer, so don't be upset if your six-month old puppy can only hold it for four hours or less.

You will want to feed them in smaller bits throughout the day rather than one big meal at once. Puppies should be eating around four times a day, and as they grow, you can cut it down until they are eating two meals a day. We will discuss food and feeding in a later chapter. It is simply important that you are making a list of everything your dog does. Even if it is something as simple as:

- 8 AM: peed outside

- 8:30 AM: ate

- 10 AM: peed inside

- 10:20 AM: peed and pooped outside

- 11 AM: ate

- 12:50 PM: peed outside

- 2 PM: pooped outside

- 3 PM: ate

- 4 PM: peed and pooped outside

- 6 PM: peed inside

- 8 PM: ate

- 10 PM: peed and pooped outside

As you create this schedule, you will notice a pattern. Then, it will be easier for you to know when it is best to crate and when you shouldn't, for example, you might crate them around 9:50 or 10 PM for a little bit as it seemed like they went to the bathroom within a shorter time interval. This is an instance where they need to be bladder trained, as they probably could have held the pee that they released inside at 10 PMuntil 10:20 when you took them out.

The best method of potty training is to simply take them outside around every two hours. Take them outside before feeding because they might not want to eat if they have to use the bathroom really badly. When they do go outside, use positive reinforcement. Take them outside as often as you can, so if you can do so every hour, that is even better. Give them a treat, and verbal praise along with a pat on their head or rub their belly so they associate going to the bathroom outside with good things.

If your dog does soil the house, do not yell at them. First, try to distract them. Maybe you call their name, clap loudly or squeak a toy. Distract them as you notice them about to go, and then take them outside as quickly as possible. If they do end up soiling your home and you discover it moments later, simply clean it up as best as you can and take them outside. If you yell at them as they are going to the bathroom, even if it is indoors, they will associate going to the bathroom as being bad, not just going to the bathroom inside as something that's bad for them.

When you see the behavior, try to make sure that you stop it. That is the only thing you can really do. This is why some people get frustrated. They want a way to punish the dog immediately in order to prevent it from happening again. The only thing you can really do is make sure you are taking them outside as often as possible and stopping them with distractions when they do want to go inside.

Clean it as best as you can to get rid of the scent. If they urinate on a certain part of the carpet, you have to do everything in your power to get this scent out or else they might go right back to it to pee once again. If any other animals have ever urinated or defecated in the house, then you'll need to make sure that this is cleaned up as best as possible as well. Dogs have amazing noses that can pick up on a lot of various scents. If you aren't careful with clean up, they will be able to pick up the scent and think that this is the spot they're supposed to go to the bathroom. There are plenty of specific cleaners that you can use which are made for the exact reason of covering up another dog's scent.

Take them on regular walks. When they go outside, give them praise. This is when you could use a clicker for training. A clicker is a handheld device that makes a clicking noise. Give them a treat and then press the clicker. When they go outside and go to the bathroom, use the clicker and give them a treat. It is a way to keep them focused on the good things that they do. They will know that the clicker means they did something good. When they go outside and hear the clicker, this means that they deserve praise. This is a method that works for many, while for some, it doesn't work at all. Nonetheless, it is a way for you to try to better teach your dog all about positive association.

After you've been taking them outside for a while, examine their stool and urine as well. Is their urine in small amounts and bright yellow? They could be dehydrated. Is it a ton of water and extremely clear? They might be drinking too much water too fast rather than sips spread throughout the day. Is their stool runny and hard to pick up? Is it hard and giving them trouble as they try to poop? These are all going to be signs that they need different foods or treats. It can seem tedious, but it will be really helpful to have a strict schedule you use to keep track of when they're going and what their stool looks like (Bovsun, 2019).

Basic Training Rules

There are a few commandments of training that are important to remember. We will discuss the important things that you need to always keep in mind not just for training, but about dogs in general. They are a different animal, and that's what many people fail to realize. We understand that they can't do certain things like drive cars, walk on two legs, and get jobs. However, too many people put human emotions on dogs. We think that they have the ability to feel guilt, reflect, learn the boundaries that humans abide by, and so on. While they can learn rules, they won't be able to emotionally grow like humans. When you make a mistake, you think to yourself, "Ah, I shouldn't have done that." You might blame other people or you might take responsibility and change for the better. When you go through that situation again, you'll know what to do to not make the same mistake. A dog doesn't have this thought process. They're mostly thinking that one thing equals bad and another thing equals good. They will simply focus on doing things that are good, things that please you because you are the leader of the pack. All they want is to earn your respect. We can't read a dog's mind, so who knows, maybe they are better at existentially reflecting and growing. However, this won't help us in training them. We have to be patient and remember that they simply don't think the same way that humans do.

Remember that your dog does not have the same sense of time that we do. They can't look at a clock and see how long it has been. Even humans don't always have the same sense of time in all situations. Think of how the ten minutes of extra sleep you get in the morning can go by in the blink of an eye but waiting in line at the DMV for 10 minutes can feel like three hours. Unless we actively monitor the clock, we don't always have a perfect sense of time. The same can be said about your dog. They don't even have the ability to look at a clock and measure time numerically. If you go to the store for an hour and a half, this can feel like the same time that it takes you to go for a ten-minute walk on your own. Dogs won't know the difference, so we have to remember that in training.

Dogs are not going to fight you back unless they feel as though their life is in danger. They can be naturally aggressive, but not because they are evil, vicious animals. They have a combination of instinct and feeling that will feed into their aggression levels. Dogs are only aggressive because they are scared. They need to learn how to play as well. If a dog is playing rough, they are simply learning the boundaries between play and real kill or attack. If they are aggressive and growing or snarling, they are likely just scared and trying to protect themselves.

If you are playing with your dog and they get to rough, make a loud and clear noise so they know that they are hurting you. Don't scream and yell, just a quick and sharp, "Stop" with an assertive over emotional tone is all that's needed. Some people will even yelp in the same way that a puppy would in an attempt to teach them about boundaries on their own level. The noise that you choose to make is up to you. Remember, you don't need to punish them. Simply stop the aggressive fight by letting them know they hurt you.

These dogs look up to you. You are their hero, their leader, the person in charge. Don't abuse this power that they've given you. Use it for good and train them to be happy and healthy dogs.

They will pay the most attention to your tone. Though they can't know exactly what you are saying, they will be able to pick up on your tone of voice to get a sense of whether or not you are happy or if you might be stressed or scared.

If you are stressed, your dog will be stressed. In fact, one study discovered that some dogs had the same cortisol levels that their human companions had. This means that they will pick up on the emotions we feel and then end up feeling it themselves.

Dogs are incredibly energetic, but they won't always be willing to run around and be animated. Remember that they need time to rest. Just because they don't want to get up to go to the bathroom doesn't mean that they're disobedient! We all know that feeling of laying down and not wanting to get up even if we do have to pee really badly. Whether your dog is a puppy or an old dog, they have the same kind of rules that apply to how you should be treating them.

Positive Reinforcement and Remaining Patient

You will want to become the pack leader for your dog, and any other pets that you might have. You can earn their trust through ways other than violence. Many individuals believe that physical punishment is the best way to teach dogs the rules. This is completely false, and studies have been proving that physical punishment over positive reinforcement actually hurts dogs in the end.

There is one important rule that you will have to live by among everything else:

<u>Never hit your dog</u>. This can't be reiterated enough. There are many people out there who might tell you otherwise. They might say, "It is their language," "It is how you become dominant," "It is how you show them who's boss," or "It is the only way they understand." Maybe it is a way to communicate. None of that matters, however. What we do know for certain is that dogs do not react well when they are hit. We know that it causes long-term damage, it physically hurts them, and they don't understand you. There are a million other ways to communicate with your dog. Screaming, hitting, kicking, and pushing your dog are not the ways that you are going to want to train them.

Know the difference between assertive and aggressive. When you hit your dog, it usually comes from an emotional place. Screaming might indicate frustration. There are emotions attached to these actions which dogs do not always understand. Separate your anger or frustration from the dog and do not take it out on them. Maybe they did poop and pee all over the house, got into the trash can, and chewed up your favorite pair of shoes. This is all terrible, but hitting them doesn't fix that, and it doesn't make them learn anything valuable. They will be scared as it is when they see that you are upset, and being aggressive toward only makes it worse.

Studies have proven that using punishment against dogs will increase anxiety and fear within them (Becker, 2012). They don't know that you are upset at them. They don't understand that you are trying to teach them a lesson. They will be confused and sad that the person they love more than anything is angry with them. Using aggression on them will make them aggressive in response to the same kind of stressful stimuli.

In fact, there is no increase in obedience in dogs who have been trained physical punishment as reinforcement for negative behavior Instead, some dogs are less obedient after experiencing repeated physical abuse. Use positivity for your training procedures. The way that they are going to learn better than any other dog is when you choose to use positive reinforcement with them. This means that rather than punishing them for the bad things they did, you'll reward them for the good things. If they are being bad, you can distract them and change their behavior by using a toy or another noise to pull their attention. If you don't catch a dog in the act of doing something disobedient, you have to wait until the next time to teach them this is wrong. If you get home from work and discover that your dog went through the garbage, they won't know that a punishment at 8:00 pm is for something they did at 6:00 pm. They have moved past this and forgotten.

Your dog should feel comfortable and trust you, not fear you. Put yourself in their paws, so to speak. Who do you feel comfortable around? Who do you love and trust the most? Who inspires you and encourages you? Who accepts you and loves you just the way you are, flaws and all? Then think of someone you are scared of, someone you cannot be yourself around, someone who makes you feel uneasy. Who is more effective in inspiring you? Which person do you appreciate more? Sure, your dog might listen to you if they are scared of you, but the bond you create with them will be built on fear, not love and trust. It is important to have a healthy and loving bond that brings the both of you together.

When training, use simple phrasing so that they can better understand what you are trying to say. Don't have complicated commands and explaining things in sentences doesn't help them. Simplify your talk. While we all love talking in long winded sentences to our dogs and they like to hear from us too, they don't really understand every single word except for relevant command words.

Ensure that you reward good behavior immediately after it happens. Don't wait too long to do this. Right after they've done something you approve of, show them immediately.

Be patient in this process. Don't overwork them. They have incredible power within their brains and can be very smart creatures. At the same time, they do have a mental limit on the day, and it is probably less than what we have. Don't expect to train them all day. Include breaks to ensure that the two of you focus on playing and relaxing together as well.

After they've started showing signs of good behavior on their own, you can reduce using treats to reward them. During the first few weeks, you'll probably need to use treats frequently to get them to sit, lay down, go outside, and so on. Once they are doing these things regularly, cut back on the treats that they have. Dogs older than eight months won't need a treat every time they go to the bathroom. Too many treats could mean that they gain weight and end up being at risk for other health conditions.

Act excited when they are behaving well instead of using treats every time. You could still give them treats every time if that's something you want. Just ensure it is a healthier treat like frozen peas or carrot sticks.

Make sure to always finish up training sessions with a positive reinforcement. Even if you trained for 30 minutes and got nothing accomplished, end it by doing something that you know they're good at, like simply sitting. This ensures that they still feel good about training and that they'll be able to pick up where you left off next time you start again.

Studies have proven that the more engaged you are with your dog, the more obedient they are likely to be. If you take your dog on your runs and walks, let them ride in the car while you running some errands, and hang out with them while watching TV, they are more likely to be obedient rather than if you leave them at home or away in their crate during these times of activity.

Specific Rules for Small Dogs

Small dogs are great companions. They have small waste, they're always by your side, and they usually love to snuggle up right next to you. They are different from other dogs, so it is important to consider the ways that we should train them especially carefully to ensure their needs are being met.

What we have to remember is that the same rules apply for these smaller dogs, but we also have to take their size into account. They might have less energy to train for longer periods of time. They might be more passive and scared in larger settings because they are so tiny, especially as puppies.

Just because a Chihuahua might not hurt you when they bite you, that doesn't mean we shouldn't stop that behavior. We have to have the same rules for our small dogs as we do our big dogs. Even if their feces may be small enough to the point that we don't care if they go in the house, we still need to discipline them when they soil a place that isn't meant for them. However, if you have bad habits in one area, it can make more bad habits in other places.

Your smaller dog might initiate a fight with a larger dog, and that's when things will get dangerous. Though you don't care if your tiny Pomeranian is aggressive, it still teaches them that it is okay to be more aggressive. Then when they run into a Doberman on the street, they park and snarl, and the bigger dog fights back. This is where you would see the negative effects of improper training or condoning aggressive and negative behavior, so don't wait until things are heated to take action.

Train small dogs to do simple tricks when you are sitting down rather than standing above them. They can be easily intimidated by size, so make yourself as small as possible when working with them to teach them new tricks.

Don't assume that you should always be picking them up either. It is easy to just lift them up and plop them outside really quick rather than harnessing them up and walking them down the stairs. However, they need to learn the path to important places on their own, so walking them through rather than carrying is the best way for them to learn.

If you have a small dog and a big dog, make sure the small dog doesn't have more privileges than the bigger dog. You might let a Mini Poodle sleep in bed with you, but the German Shepherd has to sleep on the floor. This creates imbalances in power and the small dog might think they have more control, leading to unequal treatments.

Don't assume just because you are bigger that you are the one in charge! Small dogs can be known for having the most attitude, so remember that it is important to still maintain your role as the leader of the pack.

Specific Rules for Medium Dogs

Medium dogs are usually breeds that might be family friendly. These are the dogs that the whole family can play with, hangout with, and have a bond with. Medium dogs are the standard for what we discuss throughout the book, so they don't have as many specific training needs. Rather than considering their size, it is a breed that you will want to cater to.

Specific Rules for Large Dogs

Big dogs think that they are small dogs in some situations! They don't realize how big they are. Just because they jump on you and knock you to the ground, they aren't trying to attack you. Many people associate aggression with big dogs, but it is mostly just because they aren't aware of their physical impact. A small dog might get excited and jump up on your legs when they first see you, and you might hardly notice. A big dog will do the same, but their leap could push you back or hurt your legs, automatically making them seem more aggressive. Remember, big dogs aren't intentionally being physically stronger! It is just part of who they are.

Big dogs are going to be high strung, more hyper, and of course more physically boisterous when they haven't been given proper time to exercise and play around. Since they have a lot of energy they need to expend on a daily basis, they need a lot of time to play and run around, and they need plenty of space to do this. If you can't provide them with consistent exercise, a smaller dog is a better choice.

Just because they're big doesn't mean they can't still get scared! Even if you are five feet tall and you are training a 100-pound Great Dane, they can still be very scared of you. Treat them with the same love and compassion that you would a small dog.

Bigger dogs might need more boundaries than smaller dogs, especially so that they don't scare other people. Even though you might not mind it if your Greyhound jumps up on you every time you get home, visitors, especially children, could be scared by this reaction. Always consider your dog's size and breed when training and interacting with them!

Chapter 4
Training Your Puppy Outside of Your Home

You may be able to rein in your dog at home, but in public, it is a whole other story. Not only will your puppy be more distracted, but so might you! Training should never stop, which is why this chapter will provide you with tips to ensure your dog is doing well in public. This chapter is filled with all the tips needed to not only carry out training in public, but to specifically know the best methods of training in new environments. It is healthy for your dog to socialize, so even though it might be harder in the end, you have to participate in these methods to ensure your dog is growing in the best way possible.

Going for Walks and Leash Training

Even if your dog has a big fenced-in yard to run freely in, it is still important that you train them to use a leash properly. You will be keeping them on a leash as you take it to the vet, groomer, to doggy daycare, and so on. It is better to have them know how to use a leash rather than always tugging and pulling away from you. Your dog needs to wear a collar in case it ever runs away or gets lost so that it can find its way back to you. Having a dog that knows how to be obedient on a leash also means that it will be obedient in other important areas as well.

As soon as you bring the puppy home, get them used to wearing a collar. They don't necessarily have to wear it all day throughout the house, however, they should get used to wearing it at important times such as going outdoors. They might not like it at first, but eventually, they won't even notice it or feel it. It is important to get one that is the right size and strength for your dog. This is something that you will want to spend the most money on compared to other products within this category that might not be as important. You don't want to get a cheap collar that breaks easily!

A collar is necessary to help identify the dog, should it ever get away. There will always be a level of unpredictability when you have a dog, no matter how responsible you are as an owner. Your dog might be completely relaxed as you sort through your mailbox but then sees a squirrel for the first time and runs away as fast as possible, completely shocking you as it gets away! It can happen, so a collar and leash are important no matter the temperament of your dog.

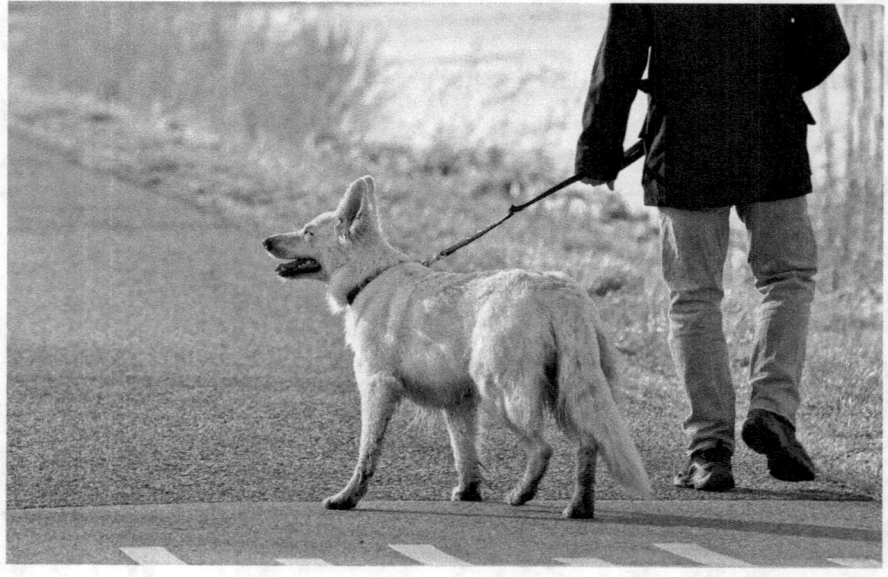

Remember, your dog will pull and tug at the leash, trying to walk as fast as possible. Retractable leashes are banned in some areas, so your choice of leash type depends on where you are located. A leash that is at least five or six feet long is good for the average person and dog. Longer leashes are better for small dogs and taller people. The leash needs to be strong enough to keep the dog back. Train your puppy in a way that teaches them not to pull the leash as hard. If you have a very large and strong dog, a chain leash might even be something that you consider.

It is certainly beneficial to use a harness while training your dog as well. This way, you can pull the dog back without hurting them. There are a few different collars that you might use aside from just a harness too. The most common is a simple collar with a plastic buckle. This snaps together comfortably for your dog.

Another popular kind is the martingale collar. This is for dogs that have the same size neck as they do their head. It will be easier to adjust these so that it doesn't slip off.

There are also half-choke collars, which are looser collars that will tighten as the dog starts to pull. This is to keep them safe while teaching them not to pull as hard.

Prong collars have pieces of metal that stick out so when the dog pulls on the leash, the prongs poke them and make them uncomfortable. This is beneficial for the owner and training purposes, but it does hurt the dog. Some dogs will be fine and stop pulling when they feel the poke. Others won't even care and will continue to tug the leash as hard as they can. Consider your dog and their level of curiosity before you use a prong collar.

You can also get a collar that wraps around their face, known as headcollars. These keep their mouths closed as to encourage them to not bark as much. It is still attached to their neck so that you aren't tugging them by their face. It is best for dogs that bark frequently.

It is helpful to have a harness connected to the collar so that the dog knows they need to heel. While you don't want to choke them out, a little discomfort can be enough to teach them how to properly walk next to you. You might just think that dogs can easily start to walk as soon as you put the leash on them. There are a few important steps when introducing walks to your dog.

Stick to one side when walking. Whether you want them to walk on your right side or your left side is up to you. However, try to pick just one that they stick to in order to make it easier for them to be obedient when you start to go for walks.

Play with them first to make sure that they are nice and tired. If you go for a long walk first thing in the morning, they will most likely be active and excited to see everything that's around them. It is better to make sure that they've been playing awhile and are more tired when it comes time to take them for walks.

It is important to start indoors so they can get a feel for what's ahead. Once you are ready to start training your puppy, put a leash and collar on them. When they are sitting next to you with their collar on, this is a good time to reward them. Give them a treat for sitting right next to you and not moving. Start walking from one point to the next. When they take the lead, stop and wait until they are sitting right next to you. Feed them a treat again.

If they pull, do not pull back hard. Simply stop so that they know to stop, make them sit, and walk ahead a bit so that you are leading. Feed them a treat and use verbal positivity when they act this way. It is okay if they get a little ahead of you, but never to the point that they are tugging on their leash. They should learn how to walk right next to your side in order to make sure they are obedient when going for walks outside.

Be happy and engaging with your dog. Don't get overly excited as this might cause them to be too excited and then they'll want to run ahead of you. Practice walking indoors as frequently as you can. Use a clicker instead of treats as you transition to walking outdoors. If they are really struggling to not pull and don't want to walk next to you, then you will have to take treats outdoors. Remind them that you have a treat so that they'll be more interested in you rather than interested in what's ahead. If you are exciting, your dog will be focused on you rather than letting their attention carry them everywhere outside.

Doggy Daycare

Doggy daycare is a great way to start to make sure that your pet is well taken care of. You can find experts that love dogs to help keep your pets happy all day long while you are at work. Not only is it helpful for you, but it is better for your puppy. They'll get a new and exciting environment, interact with other dogs, and come home tired, so that you don't have to exhaust them right away.

First, do extensive research into your doggy daycare. Make sure to read real reviews from actual people from multiple sources as well. Do your research on what they practice, what they believe, and the basic steps they take their pets on throughout the day.

Consider other amenities that will benefit your dog the most. Do they offer grooming and bathing? Do they make sure to train dogs? Do they feed them? Do they have indoor and outdoor space to run? All of which are important things to consider for your pets. They should have a thriving environment that stimulates them and fulfills their basic needs.

Ask if you can tour the place first! Many will be fine with giving you an interview opportunity. They might not show you where they keep the dog out of respect for other dogs and their owners, however, it is beneficial to examine what the kennel situation looks like for your pets.

They'll likely have a training period with your dog. This might mean that they try things out for a week to see if it is a good fit or if your dog should go elsewhere. Socialize your dog at parks and other public places first to see what their strengths and weaknesses are. Do they get excited and run up to other dogs? Does this interaction make them anxious? Aggressive? Overly excited? Keep track of how they interact so that you can help the daycare workers as well.

Check in with what foods they'll have. Do they provide food or are you going to supply it? Will they be doing feeding at all?

Make them aware of all your commands. You want to share with them the tricks your dogs know or are at least trying to learn so that they can help your pets training further. You don't want to give them two different commands for the same action because that will just confuse them.

It is a good idea to find a daycare that also does training. This way, your dog will have consistency. You can take training classes with the dog and they can have the opportunity to learn from professionals, and then you can be assured all training procedures will be the same at home and at daycare.

Make drop-offs quick and don't get too excited when you see them for the first time. if you prolong the drop off, it will make them more anxious and scared. It will make each drop-off harder and harder. Make it short and sweet. When you pick them up, they will naturally be excited. Don't make it so that they are even more out of control with their emotions. It is tempting to be like, "HI! I missed you so much!" with a screeching, excited tone, giving them lots of hugs and kisses. How could any dog parents resist? However, this will make these moments harder and harder each time because the dog will only associate positivity with you, and not with the trainers or caretakers they're with all day.

Parks

Parks are like a dog heaven. Whether it is a regular park or one that's specific for dogs, they provide a great opportunity to get some exercise while healthily socializing as well. Find a park that is close to your home which you can take your dog to regularly. It is good to find one in walking distance so that you can get more exercise for your dog before they even get there.

Check-in with all the rules allowed at the park. Are dogs allowed? Are they allowed off their leashes? You don't want to take your dog somewhere that they aren't even allowed! Remember that they will be socializing with other dogs and other people. Are they prepared for this? Do they have the skills necessary to be a present animal that doesn't attack others? Will they get excited and jump on someone? Will they bark and growl at other dogs? While it is important to socialize your pet, you also have to remember that there needs to be a level of basic training already instilled in them so you can diffuse any potentially challenging scenarios.

Consider your dog's anxiety and excitement level. Are they going to be freaking out and be overly excited? Are they going to be more relaxed and instead stay close to you?

Walk around with them before going to the park to reduce their excitement. If you take them to the park without having any physical exercise in the day yet, then they will be more excited than anything else. Don't wait until you get to the park to work them out. They need time to calm down before playing or else they might end up being far too excited.

If it is a specific dog park, then it is usually fenced in so they can be off leash. Other parks might have fences that allow you to take them off their leash as well but check in with the rules. You might also want to take them off their leash no matter what, but you have to ensure they are properly trained before ever doing this. If you ever take them off their leash with no fence, they have to know how to come on command, how to walk next to you rather than ahead of you, and their excitement with other people should be low. Even if this is the case, it is still not recommended to let them run free where there's no fence. It is not just for your sake, but for other people as well.

Don't take their leash off right away when you get to the fenced in park either. Give them a chance to sniff around. Let other dogs come up to them if they want and give them a chance to look around and get used to their surroundings. If you let them off their leash right away, they might start darting around and become concerned and fixated on the wrong things. Warm them up to this new public setting by keeping them on their leash for the first five minutes, minimum.

Don't build the excitement too much either. It is easy to say things like, "Do you want to go to the park?" in an excited tone. How could we resist seeing and showing off how cute our pups are when they get all excited. However, this is going to make them more active when you get to the park. They could be excited to the point that they're uncontrollable, and that's not what you want for your dog.

If you are taking a puppy that's not yet spayed or neutered, consider how other dogs may react. If you take your tiny Mini Poodle girl to the park and she's not fixed, she might get approached by a lot of larger male dogs. This could cause her anxiety, and she might even fight some back. Don't let this happen and keep her close if that's the case. If you have an unfixed male, he might run up to other females and try to bother them as well. It is best to have all adults fixed, but for puppies that are too young, keep them close.

Feed them long before so that they can go to the bathroom before you enter. If you feed them right before you go, they might poop right away when you get there. That's fine for dog parks, but for regular parks, this could be against the rules.

Pay attention to your dog and interact with them constantly. This shouldn't be the time to scroll on your phone. You never know what they might get into, and you will know even less what other dogs could end up doing. We can't assume that all dogs at the park have been properly trained in the first place.

It might be a good idea to leave the toys at home. A ball is usually fine, but specific toys could cause conflict. Another dog might approach your puppy, and they could get defensive over the toy. Dogs tend to be more competitive at parks. Remove the chance of conflict by leaving potentially triggering items at home.

Pay attention to what they might be picking up and sniffing at. Especially puppies. They might stumble upon a fresh pile of feces left by another dog and decided to give it a lick to see if it is a treat. The point is to have fun, so don't make it a stressful time for your dog! Be as prepared as possible to meet all their needs.

Friends' Homes and New Environments

There are a lot of people in the world who don't care about how other dogs act. They'll be excited to see a puppy and the first thing they'll want to do is shower it with attention, even if your dog might be misbehaving. They won't always be as serious about training because they aren't thinking of the dog's needs in the long run, they simply want to hold and cuddle a cute little puppy.

Though you might not care if your dog is jumping on strangers, remember that going over to someone's house might mean that they'd prefer if you didn't allow your dog to do that. You might have no boundaries for your furniture and the dogs are allowed to sit and chill whenever they want. Other people might not want your dog on their couch at all, so respect the rules of other people's homes.

You might have your friend be pet sitting as well. Whatever the situation, there are a few rules to remember. Keep them leashed when you first arrive. Just like the dog park, they are going to be overly excited, so you need to give them some time to warm up to their new surroundings. If you don't keep them leashed, they might end up running rampant around the new home to sniff out everything and anything!

Have a toy that will keep them distracted. If you are going to a friend's house to watch a movie or play a game, you'll want your dog to be focused on something other than you. Bring a bone for them to play with or another treat that will keep them busy on their own. This will help make sure that they are relaxed and free from getting into typical doggy trouble.

If they're staying the night, ensure that you have a place for them to sleep. Bring their crate with you if they are still crate training. You want to ensure the same consistency in night time routines at someone else's house that you have for your own home.

Check-in with any other pets in the house and what it might mean for all the animals. Does your friend have cats at his or her house? Has their cat interacted with dogs before? t They might need to keep the cats in a bedroom during the visit to protect both the dog and the cat.

If you are introducing it to a new dog, have your friend bring them outside. If you bring your puppy into a dog's home, this could make them defensive. Instead, have them meet on common ground so that they're more respectful of each other, and then you can bring them together.

Bring supplies in case they decide to use the bathroom in the house. While you might think you have it under control because your puppy hasn't gone inside for the past few weeks, this new environment could be the one place they decide to relieve themselves.

If you can, walk your dog through the house to the backyard or whatever the pathway might be to use the bathroom. Just like you would for your training at home, you want to show them the way to the bathroom so that they'll know exactly where to go when they need to relieve themselves.

Remember to be respectful of house rules as well. Even though you let your dog on the couch, your friend might not want that. This is why it is important to teach them 'down' commands when they're at home on the couch and bed. If they know how to get down, then you can use this command if they do end up jumping up.

What to Do When You are Not Home

You should take three or four days off when you first get a puppy. Unfortunately, there's no "new puppy" leave option for work. Imagine if we had puppy-leave like maternity-leave! After getting your dog introduced to the new home and when you have to head back to work, you'll need to use different methods to care for them when you are not home.

The first thing you should do is fill in help and reach out from friends and family to be with the dog. After six months, your dog will likely be in some healthy routines, though they're still technically a puppy. Before this time period, see if you have anyone, or a number of people, that are willing to come over and hangout or play with the puppy for a bit, or if they can at least stop in to take them outside fast. Many people will be more than willing to help out for puppies!

You can hire a dog sitter or take them to doggy daycare. There are many apps and other services that make it easy to find a dog walker in your area. They can send you pictures of the dog so that you can make sure they are being properly taken care of.

If none of these are an option, then you'll have to make sure to set the home up for them when you leave. Remember that they should never be in a crate for longer than eight hours, but this is the absolute maximum. That doesn't mean that eight hours and 26 minutes is okay. If anything, you should keep in mind that seven hours is your maximum threshold and you can go over that by a few minutes here and there, or on the days you are running late.

Take them outside before you go, but don't make it something that they are scared of. Treat it like normal. Leave your purse or other things you are taking with you inside so that they don't suspect anything. Just like with the doggy daycare, don't make them overly excited or else these will become focal points for your dog's excitement.

Wait around a few minutes after they go to the bathroom. If you take them inside immediately after, they will learn that peeing or pooping means it is time for you to leave. They might end up holding it for a while because they don't want you to leave. Then you might think that they don't have to go, so you take them back inside, only for them to have to hold it for even longer. They should have plenty of time to go to the bathroom.

Crating at first is fine, but only for shorter periods of time. Remember the month-to-hour ratio for how long they can hold their bladder. If you are going to be gone for a long time, you might consider giving them a room. This way they have several different options to lie around and play, and if they do have to go to the bathroom inside, they aren't doing it in their crate where they have to lay in it.

Tire them out if this is an option. If you can play with them before you leave for work in the morning, this makes it so that they can just focus on sleeping while you are gone. Invest in a doggy webcam so that you can watch what they're doing. Some even give you the option to press a button to talk to them or release a treat!

When you get home, the same thing goes for the doggy daycare. Don't make it too exciting or else this is how they react the first time they see anyone. Make it quick and casual and take them outside immediately. Remember that they have been laying around all day, so they will be rather energetic. After a long day of work, you may not be nearly as excited, so don't punish them because they're extra happy. Make it a point to play with them for at least 15 minutes when you get home from work, or wherever else you were. This ensures that they'll get tired out and you won't have to deal with a crazy energetic dog for the rest of the night.

Public Tips and Reminders

You might be getting a dog as a service animal. You might be someone who is more active than others. Whatever your reasoning, you will probably be taking your dog out in public frequently. We have a few more tips for you to remember when taking your dog out into the social world.

Keep your dog training even when in public. Consistency is key so you don't want to create an environment where your dog is free to do whatever they want just because you are no longer home. It will confuse them and make your training regress.

Just like the other activities we discussed, try to tire them out before going out in public. Establish some own boundaries with what you might say to those who want to pet your dog. You might not want others to distract them or make them excited, so it is perfectly fine to say, "I'm sorry, but we're in training right now," if someone asks if they can pet your dog.

Keep a few treats in your pocket so that you are not out of options should you have to reward good behavior. A clicker is a good option in public because it helps bring their focus back to where it should be without it being a distraction.

If they are acting badly, don't just give them a treat to calm them down. This will teach them that bad behavior will get rewarded. If you have one dog in training and one dog that's older and already well-behaved, try to refrain from having both of them out in public at the same time. This might make it harder to focus on training the one. Whatever you do, maintain consistency, and stick to the same rules no matter where you are.

Chapter 5

Skills that Dogs Need to Know

Though it is cute to teach your dog to sit and shake your hand, roll over or play dead, they also need to know some basic skills such as sitting and laying down. These are important things to teach your pup because it will be the way that you control them when needed as you are training. When you can teach your dog new tricks, they will be more obedient. Remember that these training processes aren't just to be done in the beginning, but all throughout your dog's life so that they're always effectively learning in the best way possible.

Sitting and Laying Down

You should start to teach your dog to sit as soon as they come home. Just because we're discussing this in chapter 5 doesn't mean you shouldn't start right away! Sitting is going to be one of the most important tricks that you will teach them. Even the most untrained dogs will still know how to properly sit. It is a natural state for them to fall into, which helps you have better control over them.

Remember positive reinforcement is key! Don't scrutinize them because they're not sitting. They won't understand why you are mad. Have some treats ready to hand out as you teach them this valuable skill.

Sitting is a great command because it is the way that you ensure your dog will be relaxed and chill when you need them to be. Maybe you are about to feed them and they keep running in circles around you. Perhaps new visitors are coming and you need them to be relaxed and sit down so they don't jump on others. You can command them to sit to ensure that they'll be as relaxed as possible in these times of great excitement.

Do these training sessions in shorter time periods. No more than 10 minutes is good. If you go for longer, it can tire them out. It can be too repetitive, and they might be desensitized, especially if they're not properly learning the different techniques.

To start, hold onto some treats. When they sit, give them a treat! It is as simple as that. You should use the normal command, 'sit,' rather than something more complicated. They might not sit on their own for a while, and that's fine. Simply give them a treat only once their furry bottoms are touching the floor and they're giving you attention.

After a few times, start to say 'sit' as they sit down. This is the way that you can associate the word with the action. Continue to tell them to 'sit' on a regular basis whenever you want them to be more composed. Each time they sit, say "good job" and give them a treat. You will discover that they know that if they sit something good will happen, so they'll do it more often. Ensure that you are also using positive reinforcement and that you have an overly happy tone.

Stick to this word because it is how others will know how to command them when you are not around.

Don't expect them to stand up again after sitting either. Step backwards or away to get them back on all fours, and then ask the command again. This is the way that you'll be able to teach them what 'sit' means rather than using your hands to force them to sit or to stand up.

Practice without the treats to make sure that they're getting it. Slowly phase out the treats after a couple weeks of this practice. They have to learn that not every time they sit they get a treat, but still use positive words so they know how good of a dog they're being.

They should learn to sit when:

- You are giving them something, like their food.

- Before you put their leash on to go outside.

- When someone arrives at your home.

- When they are being nosy and begging for your food.

- Before jumping onto beds/couches.

- Any other time you need their attention and focus.

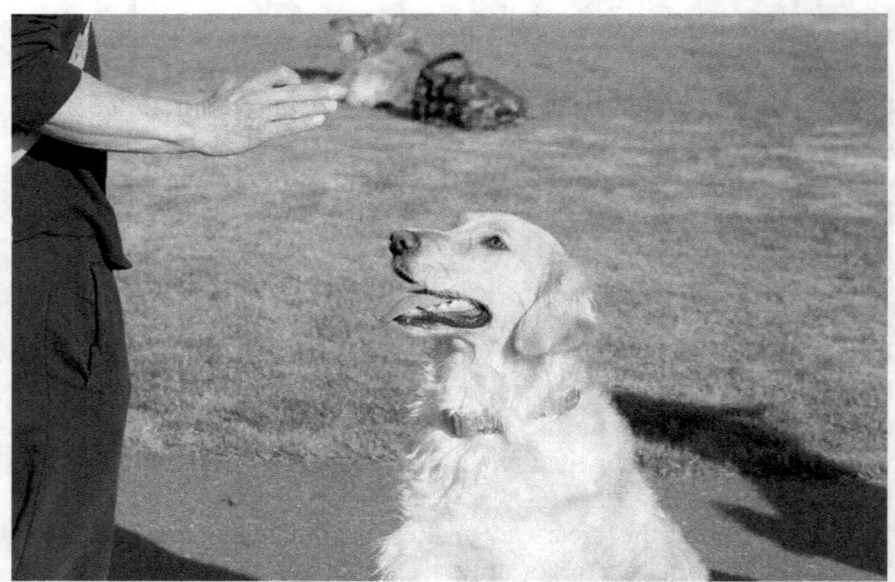

Laying down is just as easy, but they need to know how to sit first. After you've successfully taught them to sit, you will be able to teach them so many more tricks even easier. It is like learning an instrument. Once you know how to play one, the rest will come simpler. To teach them to lay down, you'll have to start by paying attention to when they are actually laying down.

You'll do it in the same way that you taught them to sit. Wait until you see them plop down and laying there and give them a treat. Once they're laying down, give them praise. They'll likely stand up from excitement, so it is a good opportunity to practice over and over again.

As you see them start to lay down, tell them, 'down.' They will then know that this word will mean that they have to lay down on the floor. Another method to teach them to lay down is to put a treat on the ground with your hand over it. they will try to get as close to it as possible, so in the process, they end up laying down. Give them the treat a few moments after this so that they understand that it is the laying down which is giving them the treat, not that this is just a random occurrence.

Never force their back down. They won't understand what you are trying to do, and it won't be the way to teach them to sit. If you are just forcing their back down, they aren't learning how to do it on their own, so they will only do this when you are the one pressing on their back.

Bark Control

Barking can be a huge issue for dogs. This is one top reason that people end up taking their dogs back to shelters or give them away as well. To make sure that your dog's barking is under control, not just for your sake but for your neighbors as well, there are a few things you can do.

Dogs that herd and have other sports jobs might be more inclined to bark. They think that barking is the way of communicating and that it is part of their job. If barking is something that's going to be a big issue for you, it is best to first not pick a dog that is known for barking.

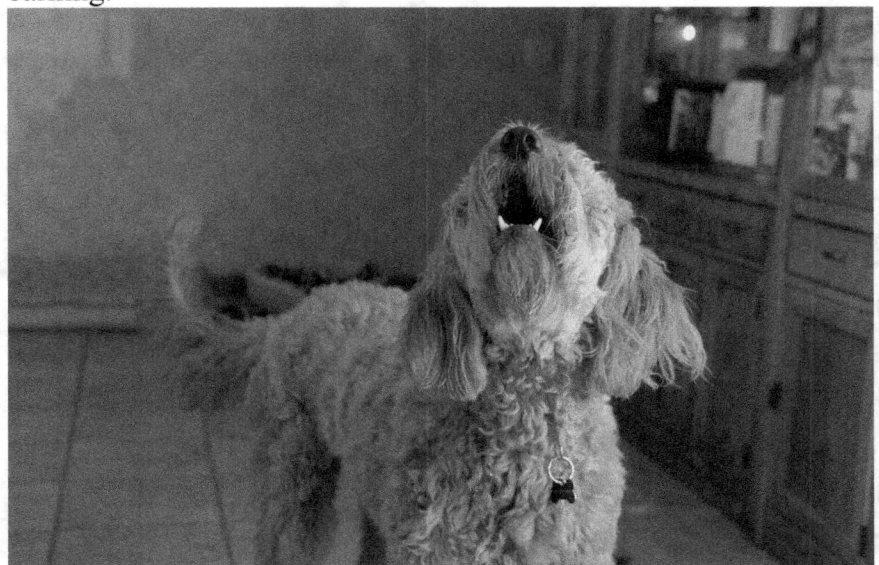

Look at the reason that they are barking so often. Is there a distraction that they always have to bark at? Maybe it is the mailman or other traffic that frequently passes by your house. Could it be that they are anxious and scared of something? Is your dog constantly barking because they think that it is playtime? Have they not been conditioned to stop barking long before this? If you look at the root of why they're barking so much, you might find a solution. For example, barking outside at neighbors walking by could be fixed if you get ticker blinds and block their view. If they're barking because they're anxious, you can do some things to alleviate those feelings.

Don't yell at them to stop barking. They will just think that you are barking with them. If they don't stop barking and all you do to stop it is add even more noise, the dogs will believe that what they're doing is right.

Wait for them to stop barking. It is frustrating but it is the best thing for them. After they've stopped barking, wait a few moments, and then give them a treat. You have to ignore them as they're barking or else they'll think about what they're doing is right. If you reward them when they're quiet, then this helps to make sure that they know quiet time is when you are the most positive.

Teaching them when to talk could actually help them be quieter. If you want to really control your dog's barking, start by teaching them to know how to 'speak.' To begin, wait for them to bark just once, and then give them a treat. Continue to do this and start to say 'speak' as they bark just as you did when you were training them to sit and to lay down. Rather than just barking whenever they want, it will help if they're trained so that they only speak on command.

This is an intermediate trick, so you probably won't really be able to do it right away, but it will still help with puppies since they can learn faster than older dogs. If training them to speak doesn't interest you, that's not the only way to have them stay quieter. Rather than yelling at them to stop, ask them to sit. When you ask them to sit, it snaps them into that mindset of needing to be more relaxed and quieter.

Remember that some dogs are just louder than others. There are certain devices that you can use if they bark a lot. You could purchase a noise maker that releases a high frequency pitch as they bark. The dog won't like the sound, so it might help them to stay quieter. You can use the clicker when they are good and click it as they are barking. It could help them to instantly snap into a calmer mindset.

Greeting Visitors and Walking Properly

As we mentioned earlier, when dogs meet new people, they will be highly excited. It is always fun to meet new people, and dogs don't have the kind of social anxiety that makes us as scared of other people! They will walk right up to anyone in most cases. The problem with this, however, is that not everyone likes dogs. Even those who are dog lovers aren't going to be as interested in having dogs jump all over them. It can be a bit too much when there is a big dog greeting you. The same goes for when you are walking down the street. A lot of people are going to be excited to see your dog and want to pet them, and your dog will notice their excitement, sometimes matching that and getting happy and jumping on them. There are a few things that can make this process easier.

Don't punish them for their enthusiasm. They are just happy to be meeting new people! They are simply excited that they are in a new environment and dogs aren't always the best at containing that kind of thrill! Rather than making them feel bad for being so happy, instead, make it your focus to simply remove them from the situation. Step back and make them sit or lay down. Once they have calmed down, give them a treat as a reward.

Continue to reward positive behavior. When they greet someone and immediately make sure to stay sitting, use positive reinforcement. Remind them that they are a good dog for not jumping on the other person.

Remember that this is a training period. You can tell your visitors this so that they are more understanding and they know that they might get caught in the cross-fire. For when visitors arrive at your home, keep the dog on a leash. Have them sit, and each time they get excited, make sure that they sit again until they are calm enough to have you let people come in. If someone knocks on the door and your dog starts getting excited and barking, make sure to not let the visitor in until they've calmed down. It is hard to not get your dog excited, especially if you are seeing someone for the first time after a while. You might want to tell them, "So and so is here!" so that they grow their excitement. Resist this so that they're as calm as possible when new people come in.

When you get home and your dog starts to jump on you, rather than stepping backwards, step forward. You want to remind them that you are in charge and you are the leader. They have to follow your lead, you aren't to follow theirs. Ignore them and don't greet them as they jump. Once they calmed down, this is when you can pet them, take them outside, give them a treat, and so on.

If they jump onto the couch and you want them down, touch the ground with a treat in your hand. Each time they jump up, do this, and have them sit on the ground. You can start to say to them, 'down,' so that they associate this instance with sitting peacefully on the floor. It will make it much easier for you and your family to tell them to not jump on them or on furniture when they have an actual command to get down.

As far as going for walks is handled, if someone comes up and asks if they can pet your dog, tell them that's fine, but make your dog sit down first. Remind the other person that they're in training and you don't want to teach them that jumping on strangers is acceptable. Having your dog know their name is going to be very important in this process as well!

Knowing Their Name and Coming on Command

Your dog's name is a very important part of their identity. This is their command. This is what they know themselves by. They associate this name with positivity, and it helps the two of you connect further.

Your dog knows to listen to their name because it means that you will be providing them with something afterwards. Calling them by their name from the beginning is important. Pick a name that is short. While Mr. Hashbrown Casserole the Third is an adorable name, you want something that is easy to call them. If you do have a long full name, that's fine, but have a nickname that they are most known for going by.

Some individuals even think giving your dog a human name will make it harder for you to treat them like a dog and not a human. This is up to you, but keep in mind that naming them after a family member or friend, or with a name that you already know might make it harder to differentiate your emotional evaluation of your dog versus this person.

Don't confuse them with a word that sounds like a command. For example, you might name them Brownie, which could sound like 'down.' Mitt might sound like 'sit' and so on. Their name should be clear and unique to them so that in all situations they always know who you are trying to refer to.

Something with a 'c' or a 'k' can help your dog to better understand if you are saying their name or something else. It is a noise that separates their name from any other noises that they might be hearing.

If you adopt them, there's a good chance they already have a name. Many breeders will refrain from name-training them. It is perfectly fine to change up their name (Keliher, n.d.). As long as you don't call them by their old name anymore, it shouldn't be an issue when training.

Looking at them and saying their name is enough. Make eye contact and say their name. give them a treat after they respond in the beginning. Don't overuse their name or else you might confuse them or teach them that they don't need to pay attention. If they aren't listening after the first few times, take a break and try again later.

As far as calling them, ensure that you start by saying the name around them. Give them a treat whenever they come to you after you've said their name. Don't make them "come here" when you are angry at them and want to use a negative tone. If you use negativity around their name, they are going to reject it and be far less likely to actually listen to you.

Don't use their name to call them when you are giving them a bath, taking them to the vet, or doing something else that they don't like. This just teaches them that they should be afraid of their name and of coming to you when called. If you aren't nice when they come to you, then they simply won't want to come to you anymore.

Fun Tricks for Your Dog

Some individuals might think it is tedious to teach dogs how to do fun tricks just for our entertainment. It is actually good for them too! It keeps them focused, and when they're obedient, they will be better behaved in all situations.

You can start with a simple 'shake.' This is a cute way for them to interact with new people. Since they do it while sitting down, they'll likely be more obedient as well. Have them sit down. You can simply wait until he puts his paw in your hand and then reward.

You could also try to tap the back of their leg to get them to put their hand up. Don't grab their paw, however. It is the same as trying to force their back down when teaching them to sit. They have to learn how to do this action on their own. It should be a natural motion. It will take some time and you have to be patient, but they will learn how to do it.

For rolling over, have them lie down. Hold a treat behind their backs and over their shoulder so that they have to roll over in order to sniff the treat. Get them to lie on their back, and then have them roll back into a lay down position. Once they completed the roll, give them a treat.

Use a treat to have them roll onto their back over and over again, never giving them the treat until they've completed the entire roll. Once they've done this, it becomes easier for them to understand what it is that they're doing which gave them the treat as a reward. Do these tricks daily in order to get them to understand the different steps they have to take to complete the trick for the reward.

Once they complete a roll, use verbal praise as well. Treats help of course, but it is your enthusiasm that is also going to make sure that they understand what they're doing is right. They will always be looking for your approval!

Having them "play dead" is super fun and will help them be the entertainment center of the party! They will love the attention. It is the same trick as rolling over, but instead of finishing the whole twist, they'll stop when lying on their back.

For this, you can associate a hand signal. Some people will hold their fingers out like they shot the dog, making it lay flat on its back. Others might ask a funny question, like, "Would you rather be a stinky cat or be dead?" This part will be up to you! It can be fun to play around with your command.

Reward them when they are in the position of playing dead. Use the treat in your hand when you make the hand gesture. This way, they'll know that the hand signal is what keeps them laying down.

Don't give them a treat right away or else they will think they have to snap out of this position quickly. Have them lie there for a moment and then give them the treat.

Practice these daily with your dog. They won't learn right away, and it will take some time for them to understand. Alternate between giving them a treat and doing it with no treat so that they do it for positive praise and not just a snack. If you give them a treat every single time they do it, and then just stop giving them treats altogether, they'll eventually stop doing the trick because it no longer means they get any treats. Don't give up with your dog, remain patient, and always remember positivity!

Chapter 6

Kinds of Exercises for Your Puppy

Playing with your dog and taking them out to the bathroom is important, but they also need to have specific moments of exercise. Just like humans, dogs can be more at risk when they aren't getting the healthy exercise needed. It could be bad for their joints, their hearts, and their bodies in general. Don't wait to exercise them! Start healthy habits from the beginning. You might discover that it actually helps you be more active as well!

Training and Agility Classes

A good method of exercise is to actually have them work with a personal trainer. If you have the means in your budget to hire someone to train your puppy, then this is going to benefit them and you. You can work alongside the trainer to learn important tricks as well as breed-specific training to help them be the best dogs they possibly can. Training isn't just for disobedient dogs – any of them could benefit! It is not even something you have to do right away either. Consider this as an option when deciding what kinds of exercises for your puppy to use.

Not everyone can afford this, so you can start to use training and agility cues at home. These are courses designed to test your dog's mental and physical skills. It is healthy for them because it provides them with exercises, and you have the ability to work with them directly, strengthening your bond. Your dog will be excited to complete tasks while also getting a fun workout that will improve his body health.

It is important that humans keep a critical mind so that they can function higher, but this isn't just true for us. Dogs with a critical mind will be more obedient. They will feel more connected to you. When they have more agility training, if you command them to do other things, they will be more likely to respond.

Agility is the kind of sport that dogs participate in where they run through hoops, fences, and other obstacles. You can purchase courses for them or make them on your own. Some individuals will use PVC pipes or pool noodles to create courses. There are common ones that you can try to recreate, or you could also come up with your own.

Your dog's safety is the most important part. Ensure any homemade courses are safe and that the dog's health and safety isn't at risk. It can help alleviate anxiety and make them more tired each day when you do consistent agility training. We just have to ensure we aren't pushing our dogs too hard when they might be feeling tired.

Start simply with some hurdles to jump over. These can be purchased at most athletic stores or you could use your own material. From there, you can set up cones for them to weave through.

To train them through these courses, you can use a simple treat in your hand to lead them through, or a longer stick with a treat at the end to help them go faster. After going through the course several times, a day for a couple of weeks, they will know how to run through with a simple command. There are many videos on YouTube that can help you deeper understanding how to train your dog for agility skills. It should be fun for them and time to play with you, not a moment where they feel stressed and scared of performing.

Swimming

Swimming will be a great exercise for your dog to participate in. It is easier on their joints and gives them something simple to do. It is a great way to cool them off in the summer as well. They will be happy to have a task, and this is another exercise that can strengthen their bond with you. They will trust you to care for them in this time. A lot of dogs love swimming, but a lot of dogs also equally HATE it. It can cause anxiety and make them feel like they don't have control. It is a different sensation for them and not every dog can handle getting wet or feeling dirty.

A life jacket is a good idea for starting with certain breeds. Most will be fine, but there are some breeds, like bulldogs or pugs, that might sink right to the bottom when they are first in water.

Don't take your dog out to deep water where there might be dangerous currents. Keep it on the shore in the beginning to make sure that their safety is entirely protected.

Make sure that they aren't drinking the water they're swimming in. Lake water can sometimes carry certain bacteria harmful for your dog, and pool water has harmful chemicals for your puppy.

Introduce them slowly. Tossing them in the water can be traumatizing. They just might never go back to water again if you do this! Ease them into it so that they can be comfortable with the water around them.

Dogs don't like to be dirty. It is good to go for a walk with them where they can have their legs be wet while still being in control of their body. If you are wanting them to go swimming in a body of water, have them walk along the coast with you. Have treats with you so that you can give them positive reinforcement when they let the water touch their feet. Don't force them in just yet and guide them along where the land meets the water. Let them sniff and pat their paw in the water. They'll be curious so encourage that in them!

If you are taking them into a pool, buy a kiddie pool first and fill it up with just a few inches of water. Throw a ball in or a different toy they like so they can retrieve it and feel the water on their feet. Let them splash around in this safe area before you pull them into the actual pool.

Give them the option to leave if they are feeling uncomfortable. You should keep them on a leash since they might even run in fear but let the leash stay loose enough, so they have plenty of space to run away if they get scared.

Never command them to go into the water and only use praise and excitement. Don't tug them into it by their collar either. It should be a natural thing that they decide to do on their own.

Once they're in the water and floating, they'll naturally try to keep themselves afloat. Don't assume all dogs can swim right away! You'll have to be there to help them stay afloat as well. Support their stomach. So that they stay close to the top of the surface. If they start to sink and their face goes under, it might scare them to the point that they freak out and try to frantically swim.

Guide them to the edge or the surface if they are scared at any point. They trust you, so don't make them break that trust by forcing them to stay down under the water! Wait until after you've gotten out of the pool and rinsed off to give any treats.

Always supervise your dog around water. Even though they might be a great swimmer after a while, they could still slip into a pool or get lost in waves from natural bodies of water.

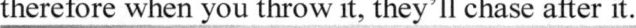

Catch

Catch is a great way to exercise your dog, one won't always require you to be running around just as fast. It is good to run with your dog, but it can also be exhausting! Playing catch gives you the chance to relax and enjoy the outdoors while your pup gets plenty of exercise.

A lot of dogs will naturally catch. It is in their biology to want to retrieve things for their leaders. They know that you enjoy the ball, therefore when you throw it, they'll chase after it.

Start with a toy that they will actually want to play with. Not all dogs are that interested in balls. If they prefer a rope or other rubber toy, throw this for them. After you've done this a few times, you could transition to a ball.

Reward them as soon as they start to play with the toy. Use verbal praise to remind them that you are excited and that what they're doing is right.

Put it on the ground in front of you and encourage them to play. This primes them so that when they catch the ball, they bring it back to you rather than run away with it. Once they bring it back to you, you can then give them another treat.

This is a good time to teach them to "drop it" too. Many dogs will bring the ball back, but they won't always let it go. They can't eat a treat with a ball in their mouth, so you can tell them to "drop it" and when they do, give them a treat. As they're eating you can take the ball from them. This is a great method to help them learn this trick. Next time they pick up something they're not supposed to as well, you can tell them to "drop it" and they'll know what this command is.

Dog-Led Walks

Walking can be so routine for dogs. It is a way that they go to the bathroom, get a reward, then go back inside. Most of the time, they take your lead. As a fun exercise to do with your dog, let them be the ones to decide which way you are going to go. Some individuals believe that you shouldn't let the dog lead because this is destroying your role as the alpha. That is debatable, so it is up to you to decide if this is a method you want to incorporate or not.

Every once in a while, you can still let them take the lead. Don't let them pull on it, but if they want to go to the left or the right, backwards or forwards, let them be the ones to decide. Keep a loose leash so that there isn't as much strain as they walk around.

This is best in nature areas rather than structured streets where they might be passing a bunch of other people. There will be fewer distractions so it will be more about them exploring their surroundings rather than getting excited about new people.

Go somewhere that you could take multiple paths. Give them options of how they can explore. It is fun and relaxing for your dog. It makes them feel natural and like they are accomplishing something.

Creating a Routine

The most important thing to do for your dog is to create a routine for it. Dogs need to be able to stick to a schedule. It gives them assurance that they'll be safe and protected.

Make sure that you are feeding at the same time every day. If they have a random feeding schedule, it will be stressful for them because they'll never really know whether or not they will have food to eat. They'll also worry that they won't be able to eat when they're hungry.

They will go to the bathroom around the same time as well. It should be a routine for them to wake up and go outside, go before bed, and a few similar timeframes throughout the day.

Sleep is essential for them. Some dogs can get anxious and not be able to fully sleep. If they don't have a schedule, they might not be able to become relaxed enough to get that deep sleep needed.

Most importantly, have an exercise routine. Not only will it be good for your pup, but you'll benefit from these periods of high activity as well. They will know the difference between play time and relax if you make sure that you are carving out specific times for them to exercise. Rather than having an active pup that acts all wild, you can have a dog that knows when it is time to lay down and when it is time to be crazy and play.

Make sure that you are also conscious of the methods that they might be associating place and specific locations with parts of their routine. For example, ensure that they have a proper place to sleep every night and that it is consistent. They will know when you take them to this place, that it is time to sleep. Have them eat in a similar place as well. If they are led to their food bowl, they'll know that it is time to eat.

Not every day is going to have the same schedule, but when you are doing those routine things, they should have the same structure each time. Keep procedures consistent so that there's no confusion on your dog's end.

Taking them to the bathroom first thing in the morning is a good habit. At the same time, it is ok to wait just a few minutes if you need to set up the coffee or go to the bathroom first yourself. You don't want to rush and stress your dog out with that first morning bathroom trip. Give yourself enough time so that this period can be calm and relaxing for your dog.

15 minutes a day is the minimum playtime you should have for ALL dogs. The bigger the dog, the more exercise they will need. If they have a day where they don't get out much, make sure that they're twice as active the next day.

Chapter 7
What to Do for the Health of Your Dog

Your dog has plenty of health needs just as humans do. Choose a vet in your area that you trust who also has good reviews. Though you might be able to train and exercise them properly, it is their health that is still important at the end of the day. Ensure they live a long and happy life by always checking in with their health.

Initial Shots

It is important that we take care of our pet's health by giving them regular vet checkups. Going to the vet can be expensive, but it is something that needs to be done. If you can't afford regular vet check-ups, it is probably best to wait until you have some money saved up to adopt a dog. When you do get your puppy, you should have a savings account with around $250 - $1,000 in it for any pet emergencies. You don't want your dog to get a random sickness or a serious injury that you wouldn't be able to afford to help them out in the end!

Many dogs will already have necessary shots if you adopt them from a shelter. This is a great option for those who can't afford all the shots up front. It is more convenient for you as well, so consider this before you adopt from a breeder.

Having vaccinations are also important so that when your dog is exposed to other animals, it can't catch, or spread any diseases. Puppies don't have the ability to get all the shots they need right away, so if your adult dog isn't vaccinated, it could affect your puppy. Aside from a few rare cases of having various reactions, getting your dog vaccinated isn't going to have any negative side effects.

Some laws require mandatory vaccinations. Check in with your local vet to see which ones you have to get in order to abide by your area's laws.

There are certain core vaccines that your dog will have to get. These are common health issues among dogs that you'll want to protect your pup from ever having to get. According to the ASPCA, these are the illnesses and diseases that core vaccinations will prevent:

- Canine parvovirus

- Distemper

- Canine hepatitis

- Rabies

There are other vaccinations that might be necessary for your dog. These could be things such as:

- Bordetella bronchiseptica

- Borrelia burgdorferi

- Leptospira bacteria

For puppies, the vaccinations should occur within six to eight weeks of their life. They will only get a few at a time with three-week periods in between (ASPCA, n.d.). By 6 months, they should have all of their proper vaccinations. Check-in with your vet to see if there are any that they'll have to come back for as they age.

It will be up to your vet and your dog's needs to determine what vaccinations are required. Its best to do as much as possible because the only downside is that they might cost more money. These vaccinations will often be a one-time thing, so it is not like you'll have to continually save up more money to keep up with regularly purchasing them.

Don't administer vaccinations on training days. They're going to be tired, and likely stressed out from going to the vet already! Give your dog a chance to rest and heal. The biggest issue you'll notice is that they are tired and that the injection site might be tender.

Anything too concerning, contact your vet. For the most part, however, they'll be completely fine and safe from getting anything serious.

Spay and Neutering

Not everyone will choose to spay and neuter their dog. You might use them to hunt or breed, so instead you'd prefer to keep all their reproductive systems intact. This is up to you, but unless you need their reproductive organs for a specific reason, you should be getting them spayed or neutered.

Spaying is when female dog's organs in their reproductive system are removed. Neutering is for males. For females, the ovaries and uterus are typically removed. For males, their testicles are removed. The actual penis and vulva are not removed on dogs. Emotionally speaking, it is a hard process for your dog to undergo, but in the end, they will be happier and healthier because of it. They might have a more relaxed temperament, gain weight easier, and be calmer in general.

It is important first to ensure that they don't reproduce. Having one puppy is hard enough, so you don't want to have more to take care of at once! After around six months, they will be able to start reproducing, so this is the age that they're usually spayed or neutered. Some studies have actually shown that waiting a year to spay or neuter larger breeds, like German Shepherds, is actually better for their health to give them the full chance to develop first (Dayton, 2016). Whatever you do, try to spay or neuter your dog between six months – 18 months of their life.

For female dogs, you will cut out their heat cycle. Some female dogs might even bleed when they are going through their heat cycles. It doesn't do much to them other than cause uncomfortability. They will likely be licking that area more when they're in heat, so they could end up causing sores if they do it too frequently. The bleeding could happen throughout the day as well, so it could end up staining your furniture or wherever else the dog might sit.

For male dogs, they might resist the urge to 'hump' things. Unfixed male dogs will usually do this to other animals as well as many toys. Even after they're fixed they might still do this, but it will be less frequent. It is usually not a reproductive issue and something that you have to train them to stop doing. Catch your dog before they start to hump. As you see them mounting something, command them to sit or lay down. Once they've successfully done this, reward them with a treat.

Both female and male dogs will also be more likely to listen, and it will be easier to train them. They will come on command easier and you won't have to worry about them being disobedient.

It can drastically reduce the risk of certain cancers or infections, such as pyometra. The lifespan of dogs that are fixed versus those that aren't is also longer.

You might be weary because you worry that it is not a natural process. It is also not natural to deprive them of these urges. It can end up hurting them in the end. They will still seek out other dogs to reproduce with and they might get stressed because they aren't able to find a mate. If they are fixed they will naturally be calmer, and they won't think about this as frequently.

Keep them away from other animals once you have gotten them fixed. They might be extra sensitive so you don't want to cause a fight that could make their stitches become loose.

Reduce their play time for the first few weeks after getting them fixed as well. You don't want to push them to overdo it or else they might end up hurting themselves in the process.

Give them two weeks to recover. This is the usual time period it takes for them to fully heal. They are getting internal surgery, so it can be an exhausting act to recover from.

A cone is probably going to be necessary because they'll want to lick their wounds. Make sure to keep this on them at all times that they are unsupervised, and even when you are together. It could only take one lick or bite to remove stitches, which could mean having to go back to the vet and do the whole stitching process over again.

Picking the Right Food and Treats

As you can see by now, the best method of training your dog is to use positive reinforcement and lots and lots of treats. For both food and treats, there are a few health guidelines that you should try to follow. Dogs aren't like us in that they won't care about tasty food as much. We have over 9,000 taste buds while dogs have far less, estimated at 1,700. They can still taste, it is just not as important to them. Many humans don't realize this and will often put their human emotions onto dogs. They think that these animals deserve to eat complex meals just as we enjoy, but this doesn't matter as much. We should be providing them with good and healthy food, but they aren't going to need to eat a steak dinner every time you do, so don't feel guilty that their food isn't as tasty. It is the smell of this food that they will care about more than anything (Finlay, 2017).

Dogs are also omnivores. Unlike cats, they should have some fruits or veggies in their diet. Raw veggies, like carrots, frozen peas, bananas, and apples are all great treat substitutes. Since you will be training them so frequently in the beginning and providing them with a ton of treats, you will want to consider these natural options to keep their weight under control.

The food you dog consumes will directly affect how they behave and how long they might live. We want our furry friends as healthy as possible so they can live a long time by our side! Choose the right food with added benefits and natural substances in order to ensure their health is in tip top shape.

Foods that are cheap and the popular brands that you can find at any convenience store will more than likely not be the first choice. Many dried foods are filled with carbohydrates that just add calories. Yes, dogs have to watch their carbs and calories too! Don't feed your dog only junk food. Many popular brands are equivalent to humans eating chips and other junk food as their sole meal.

The first three ingredients are the most important. Check any bag of food and look at the things that are listed before everything else. If these are chemicals that you can't even read, you might not want to have this be something that you give your dog.

The worst ingredients include:

- Meat by products (typically carcass or other animal parts that aren't actually meat)

- Caramel color (no dyes or additives should be in dog food)

- Oil (oil is dangerous for dogs and can cause inflammation)

- BHA or BHT (preservatives dangerous to health)

- Animal fat

The best ingredients will be:

- Meat

- Grains

- Vegetables

- Vitamins, minerals, and nutrients

- Meat meal

Your dog should be eating foods that have AAFCO approval. This means that it has actually been tested on dogs. Not all brands will give their food to dogs before even releasing it. Though it might taste good, it might not have any actual benefits for them.

Puppy specific food should be used for younger dogs. They have different nutritional needs that have to be attended to as they are still growing and developing. It will often be softer as well. Consider their breed, too. Some small dogs will have different food they need to eat, and larger dogs might have various needs as well. The tenderness and the size is important.

Wet food may be a better choice. The food that they eat in the wild is wet and moist, not crunchy bits all the time; however, you might like dry food because they don't eat it all at once. Check with the package to see what the serving size is. For puppies you will want to feed them three or four times a day. Once they're a year old, cut it back to just twice a day.

Water should be out all day long. If you have multiple pets, you should have multiple water bowls. It will be a good idea to have one more water bowl than people that you have. Never limit their access to water and ensure that no matter where you go, they will have the chance to drink as much as they want.

Avoiding People Food

It is hard to not give your dog a little snack when you are eating. You want them to like you and when you are enjoying food, you might want them to get the chance to enjoy this just as well! However, if you are giving them too many table scraps, it will create bad habits.

Not all people food is harmful to dogs, but you should avoid giving them your food when you are eating. It will create negative boundaries with your dog. It is a good idea to keep them in their kennel when you are eating so that they aren't trying to bother you too much while you have food in front of you. You should feed them after you eat, as this indicates to them that you are the leader of the pack, since in the wild, the leaders get to eat first.

Avoid giving them any human meat because there's so many factors that could be harmful. Even if you get it at a restaurant and want to give them scraps, you never know what they might have seasoned it with initially.

Fruits and veggies are fine snacks. Give them to your puppy as you wish but remember to not do it in the space that you eat. They should know to respect your boundaries and leave you alone as you try to eat your meals. These healthy fruits and veggies include:

- Carrots

- Bananas

- Apples

- Watermelon

- Cucumbers

- Peas

- Sweet potatoes

- Celery

These are great treat alternatives. If you don't want to buy a ton of treats all the time or find that they cause a ton of weight gain, then switching to these foods is a better option.

Make sure that anything you give them isn't seasoned. Dogs can't have garlic and onion, among other flavors, so you want to ensure that there are no hidden seasonings on the snacks you might be giving them.

Make sure that you never give them scraps directly at the table. They will know that it is fine to beg. Take the scraps to where they eat. Create a habit of making them sit or lay down in a different room when you are feeding them. You might think it is fine to do this at your home, but when you go to a friend's home or have entertaining guests over some meals, you don't want your dog to constantly bug them.

Ask yourself if your dog would find it out in the wild to eat on their own. They might naturally find certain fruits and veggies, but popcorn, chips, and bread they wouldn't. The key is to give them things that are going to help their health, not just satisfy their taste buds. If your dog doesn't eat their dog food, then you'll have to consider if they simply don't like it and purchase a new brand.

Flea Treatments and Bathing

Fleas are annoying, but they won't be the worst thing that can happen and are usually easily treatable. The worst thing that you will find if you get fleas in your home is that you get bitten and your bites might be itchy and swollen. This is treatable with some Neosporin ointment or other anti-itch cream to help your bug bites. It is worse for your animals, however, as the fleas can actually feed, and breed off them. Not only will the fleas bug your pets, but they will scratch, and this could cause even bigger sores and tender spots on their bodies.

They will get trapped in their fur and it can cause your dog to aggressively itch themselves. We scratch our wounds as humans, but we usually know when we need to stop before it gets bloody. Your dog doesn't know as well and will scratch and scratch and scratch. This can make their wounds even bigger and cause infections. If you notice your dog frequently scratching, it might be a sign that they do have some sort of insect issue.

Though fleas are more treatable than other issues, don't pretend like they're no big deal, because they can still carry viruses. Some fleas can go from two to 2,000 within a couple of weeks!

You can look for fleas yourself by separating their fur and looking at the root of the hair. On their skin you will be able to see the flea crawling around, or you will notice the flea dirt.

When you do see a flea, the only way to ensure it dies is to drown it, as squishing them doesn't always work. They have a hard-outer shell, so it is often the best choice to flush them or rinse them down the drain with soap and rubbing alcohol. If you do find a bug and it is not very tough to squish and kill, then it is probably not actually a flea and instead some other sort of insect.

You can give your dog a flea bath and then give it a flea treatment. The first bath will be to make sure that as many fleas and flea dirt is washed away as possible. You'll then want to give them a flea treatment. The best kind is an ointment you put on the back of their neck. This soaks into their natural bodily oils and will eventually spread throughout their entire body. About a half a week to an entire seven days from giving the flea treatment, you could do another bath to get rid of any fleas that might have died. Check the box of the flea treatment you purchase to see if you need to wait longer.

Treat your dog not just for its own sake but for other pets around the house as well. Your dog that goes outdoors might bring home fleas to your indoor cats. If you have indoor cats and a dog, you will still want to give them treatments. The dog might spread fleas to them, and then the fleas could spread back to your dogs. It is best if simply all cats have been treated.

If your home seems to have fleas pretty bad, you can do a home treatment. This is usually in the form of a smoke bomb that will spread throughout the entire house. Ensure that no animals or humans are home when you do these treatments.

As far as regular bathing goes, this will vary depending on the dog. Some dogs will need regular bathing while other dogs will be perfectly fine going a year or more without a bath! It will depend on the dog breed whether or not it needs bathed.

Short-haired dogs will need bathed less. Make sure that you regularly brush all dogs, short or long haired. They will still need brushed in order to remove dead skin and loose hairs. It will help regulate the hair they leave behind on your furniture as well. Outdoor dogs that are very active need more bathing as well.

Once a season, or around three-or-four-month intervals is usually the minimum for bathing. This would be for bigger short-hair dogs who don't need as much grooming. The most frequent you should do is every other week. This would be for dogs that need regular grooming and whose hair grows long.

Keep their hair groomed so that it doesn't get too long between their toes or too long over their eyes. If you live in a hot area, it is best to maintain a buzz cut so that they don't get too hot. When it is cold outside, having a thick coat will help them stay warmer and make it easier to encourage them to go outside.

When bathing, pick only a dog-shampoo. Sometimes people will use baby soap, but there might still be dyes or perfume in this that could be fine for a human but not for a dog. Make the water warm, and have the environment be relaxing. Try to ease your puppy into the bathing process by making it exciting and giving them treats. Some people have actually used the trick of spreading peanut butter on the bathtub wall so that the dog licks it while they're getting bathed!

Chapter 8

What Not to Do When Training Dogs

There are some common mistakes in dog training that you'll want to do your best to avoid. There are many dog owners in the world that might not give their puppies the best chance when training, and then they end up wondering why their dog isn't behaving the way that they want it to. To make sure that your dog is going to be properly trained, we've created a list of what not to do so that you don't make any of the common mistakes that other pet parents might!

Aggression in Training

As we already discussed, aggression will just make them react more aggressively in the end. Though they might get frustrating at times when they don't listen, and they might have times where they don't have any progress in training, using aggression on them can cause them to be afraid and will actually regress any valuable training you did use. These are the most common forms of aggression and you shouldn't be using it on your dogs ever:

- Hitting

- Kicking

- Forceful pushing

- Threatening body stance

- Verbal aggression

- Yelling

- Pointing

- Other forms of physical and verbal violence

Some dogs are rather aggressive, so you might ask yourself, "What am I supposed to do to control my dog?" You might have a 120-pound dog that never seems to listen when you are telling it to sit down. Perhaps you have a little poodle that never stops using the bathroom in the house. Though these are frustrating situations, violence isn't going to help. We're going to discuss a few ways that you can assert power that you don't need to use any violence on them.

Not only does hitting your dog hurt them physically, but it destroys that bond. You are the one person that they trust more than anyone. How are they supposed to trust you if you hurt them so much? They won't be able to listen to you because they're scared of you. The more aggressive and angrier you are with them, the more that they will have trouble following your lead. They will grow their fear, not their trust, and that's a quick way to kill anything the two of you have built together.

Let's first understand the reason that they're aggressive. Dogs are very complex creatures and they have some needs that have to be met, while also having different stressors and anxieties just as humans do. Though they don't feel the same way that we do, they can still experience some of the same effects of stress.

The first reason is because they are trying to be protective. Dogs are territorial animals, like many other creatures. They want to be able to help protect your home. They care about you and want to ensure that you are protected. If they are often aggressive towards others, they are likely trying to protect their turf.

They are often frustrated and feel the need to gain control in some way. Just like how some people are more easily flustered, many dogs can be confused and turn angry too. They might be overwhelmed by others and feel the need to be aggressive to try and gain control. When you don't know what's going on, you might find that you panic, and this is how some dogs will operate.

They might be aggressive because they haven't been socialized. This is why it is important that we start to socialize our puppies from the moment that they're brought home. If we don't, we could end up making them scared of visitors and more likely to lash out on others.

Animals have a fight or flight instinct just like humans, so showing signs of being aggressive might simply be because they are anxious. In order to assert dominance, there are a few things you can do.

Be consistent with the boundaries you set. You can't have rules that apply during one part of the day, but not the other. You might not care if they jump up on you, but you have to set this boundary so that they don't jump up on strangers.

Sometimes, forcing them to be affectionate can actually damage that bond. If you are constantly squeezing and kissing them, or even holding larger dogs that don't like to be picked up, it could make them more fearful. They might not know what you are doing, so let the dog come to you when you want to show your affection. Be gentle and patient with them. If they are sitting next to you and licking you, you could try to put your arm around them and be affectionate, but if they're running around, eating, or playing, don't try to force a hug.

You have to have confidence over them. They will smell fear. If you are stressed, it might make you stress more, and then you could end up releasing chemicals like adrenaline through your glands. Dogs can smell this and know that something is wrong. If you are frustrated and stressed out when training them, they will sense it and it will be harder for them to focus.

Never stop playing with them! Get their energy out through playtime, not through times when they are rough and constantly jumping on others.

Feed them after you eat your own meal. As we already discussed, this is common for the leader of the pack. If you want your dog to really listen to you and remain obedient to what you are asking them, you have to ensure you are establishing dominance over them in a way they respect. You can also take them for a walk right before they eat so that they know you are in charge. You are making them work for their food!

Inconsistencies

The biggest killer for your dog's training is not being consistent throughout the process. Dogs need routines and they only know what boundaries are in place by experiencing them on a repeated basis. If you are consistently inconsistent with your dog and the rules are always changing, they won't know what they should be doing.

What you do at home is what you have to do everywhere else. Even if it might be hard to remember, always keep your dog in training. It will be a long process that will take time to get used to. They need to follow certain boundaries in order to stick to the rules which you are teaching them.

Their behavior should be right or wrong at the core. It is wrong to jump up on people. They won't know that it is wrong to jump up on strangers, but okay to jump up on you. They will simply know that it is fine if they do jump.

Even when you don't feel like doing something, you have to stay consistent for your dog. It can be hard to always take them out at the same time, and there are moments where you might just want to stay in bed a little longer even though your dog's already woken you up to go to the bathroom. Even in these moments of struggle, it is essential that you are taking care of your dog's needs and staying stable.

Different partners should not have different rules. This is why it is so important to have a family meeting to establish all of the rules. Even if you have a roommate who's not as interested in caring for the dog, they have to help out just to ensure that no bad habits are created.

Dogs don't have a way to communicate with us other than through signs or symbols. They won't be able to see these if they aren't consistent and presently clear. They need that regulation to understand, as it is really the only true way that they can communicate with you.

Useless Repetition

Your dog needs routine and consistency. However, if you are consistently doing something that's bad for your dog or that they don't understand, then this isn't an ideal routine that will help them in the end.

Useless repetition is like calling their name over and over when they clearly aren't coming. It is like trying to make them sit by pushing down their butts over and over, never able to actually do it on their own. When you notice that you are repeating something to your dog over and over again and they don't get it, then it can be frustrating. However, there are certain steps that you'll want to take instead to get them to really understand what you are trying to teach.

For example, you might find that you are trying to teach them to roll over. You've done it twenty times now and they still don't get it. The first thing that you'll want to do is ensure that you take a break. They might simply be tired, and you could be showing frustration, a bad recipe for pointless training.

Your emotion will change throughout the useless repetition. At first you might be excited with a happy tone while you are training your dog. By the end, your voice might shift and you could end up sounding frustrated. This will only confuse the dog further as they won't really understand what it is that they did wrong to make you frustrated.

Grabbing his paw and shaking it over and over again isn't going to magically work on the 50th time when you are trying to teach him to shake his paw. When something isn't working, you have to reevaluate the process to find a method that they actually do want to participate in.

Sometimes, the repetition is the way that they end up learning. They think that you have to get angry and say 'sit' five times over and over again until they actually decide to sit. Remember that your dog won't think like you do, so they aren't going to understand the pattern you want if you are trying to teach it in an unhelpful way.

Confusing the Dog

It'd be so much easier to train dogs if we knew how to really talk to them. Unfortunately, human-dog communication is still limited and rather than finding methods to speak to them, we have to discover the way that they will understand our signals. Let's discuss what you might be doing wrong which could confuse them. If the dog doesn't understand what you are trying to do, it can make them upset and then it will be harder to train them in the end.

The first thing that confuses them is when you punish them for natural things. This includes things like going to the bathroom, eating, and playing. If you notice that they are going to the bathroom in the house and scream and yell at them, they will think that they're getting in trouble for simply going to the bathroom. If you get home and see that they are eating food you had sitting on the counter, they will think that they are in trouble for eating. You have to catch their bad behavior before it happens, or right as it starts. Punishing them in the middle of it will simply be confusing you.

Did they step on your foot and you yelled? You are not likely to be able to train your dog to not make mistakes. They will still stand behind you when you are cooking sometimes, or they might jump onto your bed and accidentally stick a paw in your gut. Though these accidents are annoying, you can't punish your dog for this because they won't really understand what it is that they're getting in trouble for.

Did they have diarrhea on the floor? They're sick, not intentionally being bad. Though it is bad for them to go inside, they won't be able to hold it all the time, especially if they are ill. The same can be said about them vomiting. Though they might be potty-trained, there are still some accidents that are entirely out of you and your dog's control.

Having the wrong labels for certain cues can make it hard for them to know what to do. Check and ensure that you are not training them to do a certain command with several different words. For example, you might use "sit," "down," or "sit down" all just for the same command for them to sit. Though they all mean the same thing, your dog doesn't know this, and they could end up getting confused.

Try to refrain from using your finger and hand to "punish" them. They will really only understand your use of different words and symbols, not necessarily hand gestures.

Make sure that the affection you are giving them is at the right time. Even though you might want to hold and snuggle them all day long, if they don't seem into it, you can't force it. Your dog might end up thinking that this is some form of punishment, but they won't be sure why you are punishing them in a certain way.

Pay attention to the eye contact that you are giving your dog. They take a lot into consideration when evaluating your eye contact. If you are staring for too long it could make them nervous or anxious. They might even end up getting aggressive depending on the context, thinking that you could be challenging them.

Abandoning a Ritual or Habit

Dogs will need trained throughout their entire puppy-life, and into adulthood. You can't stop training your puppy just because they seem to know the same old tricks. As they get older, you won't have to train with them so often, but there are still important measures to take to ensure they keep up with their knowledge on certain tricks. You might train your dog to play dead within the first few months of their life. If you stop doing this trick, by the time they're two years old, you can't expect them to immediately remember how to do it.

Just because you started once doesn't mean that they will know this forever. In order to ensure that they are keeping up with the tricks that you taught them, have moments where they have to do the treat in order to get a certain reward. The best time to give them trick refreshers is right before they eat, before you take them outside, or after you come back in from taking them out. Give them a treat and positive reinforcement. It only takes them doing these tricks once every couple of days for them to maintain their knowledge of a certain command.

Sometimes, you might abandon an entire ritual. Perhaps you no longer take them outside first thing in the morning and instead wait until it is convenient for you. This can mess up your dog's entire schedule and make them rather confused! If you do need to change protocol, remember to transition your dog. They are already used to a certain way of life and become easily confused. The sudden change could stress them out, so do your best to keep your dog as calm as possible.

Treating All Dogs, the Same

All dogs are different. Even though they have similar needs and will react in usual ways, they're also very different from one another. Some dogs are lazier, some are more aggressive, some are very needy, and others might be independent. Whatever you might have thought about dogs before could be completely wrong after meeting that one special dog that just isn't like any others!

Just because you trained three dogs already doesn't mean this fourth is going to be the same. A mistake that some will make is thinking that all dogs act the same way. They might believe that if a dog isn't acting in the same way that another dog has in the past, that this new dog has something wrong with it. You might have just had a dog that was easier to train first, so the second dog isn't necessarily bad, it is just a little more difficult this time around.

Some dogs will be naturally shyer than others, while some may be boisterous and high energy. Others might simply be ornerier. Some dogs are excited and active, others are scared and want to sit away from everything. Some dogs love new people, some dogs get scared of strangers. Some dogs bark at cars, some dogs like riding in them. Some dogs will play with 100 different toys, some dogs might only want to play with one. Some dogs will be excited to see you, other dogs might not care as much. Each and every dog is different. Not only might they be born that way, but how they were raised before meeting you will also affect their development.

Don't punish your dog because it is not the same as others that you've had. Respect their individualities just as we do for humans! You can assume that some things are going to be similar, especially if that helps in your training. At the same time, remember that some of the things about their personality aren't bad, just different!

Positive Reinforcement with Bad Behavior

One of the biggest mistakes that new trainers can make is that they will end up using positive reinforcement for certain bad behaviors that their dogs have. This occurs when a dog is doing something bad, and the owner gives them a treat in order to stop this bad behavior. In the process, however, the dog learns that if they do this bad thing, they'll get a treat in return.

Sometimes you just want to get your dog to shut up! You just want them to stop barking, maybe they're scratching, or simply being violent. While giving them a treat is a great way to stop them, it also tells them that if they just bark enough, if they get aggressive, or if they start to chew something up, they'll get a reward in return.

Don't use a reward to get them to stop doing something. Instead, use a common command to get them to stop, such as 'sit' or "lay down." Once they have done this good thing, then give them the treat. Next time the dog is barking at cars and you can't get them to stop, don't call them over to give them a treat. Call them over, make them lay down, and then give them the treat. They will better understand that the only way to receive a reward is to be calm and relaxed.

Your dog is smart enough to know how to get what they want. This is just as we mentioned previously with crating the puppy. If they whine and then you let them out, they will know that whining means they can get their way.

Positive reinforcement in negative situations might confuse them as well. For example, let's say that your dog is really afraid of thunder. Whenever there's a storm, they go and hide under a desk. Every time you go to him and say, "it is ok, don't be scared," over and over while you pet him. The petting can help, but they will also pick up on your words. Next time that something scary happens and you tell them, "it is ok, don't be scared," they might be brought back to those same frightened feelings that they had the first time.

Not Proofing Tricks

If you train your dog to sit in the living room within the first three days, that's wonderful! The next step is to 'proof' this behavior. This means that you are taking a trick that they learned and doing it in different settings with various distractions to make sure that they understand the trick at the core. If you teach them a trick in one room, they might only thing that they should be doing that trick in that room. You have to repeatedly do this everywhere so that they understand what the command means at the core.
Dogs don't have the ability to make generalizations like we do. They're not going to understand that just because they learned to 'sit' in the kitchen they're supposed to sit in the living room as well. To really make sure that they're understanding this, you have to proof your tricks.
Start to do the same tricks in different areas throughout your home. Make sure that you do it in the backyard, on walks, and in all rooms.
Add in other distractions after this as well. Do it at the dog park where others are watching. Do it when you have their food prepped and ready to give them in their bowl. Create methods of excitement so that they understand they should be present in all moments and focusing on you.
Do it without a treat in between sessions. You don't want them to stop doing the trick just because you stopped giving them treats. You might have to teach your dog the trick over again, but this is the chance for you to strengthen their abilities. Patience is going to be key! Don't forget to proof or else your hard training work will be lost. If you are giving them treats and showering them with positivity, then they are understanding that this anxiety is fine to have in these scenarios. While it is not a bad thing for them to be scared, don't use positive reinforcement or else they will react the same way next time.

The best thing you can do to comfort your dog is to distract them. Get their favorite toy and play with them. Give them commands and tricks, rewarding that behavior with treats. Act normally and continue to be excited, not sad or scared. Don't force them either. Sometimes it is best to leave them alone while still being in the same room just so that they know you are there to keep them safe. You might put your arm around them or pet them, but refrain from using positive language or treats because they will get the idea this is how they should act in fearful situations, making them only more scared in the end.

Waiting Too Long

If you wait too long to start to train your dog, you are wasting precious time that could be spent helping them grow. Old dogs can learn new tricks, but it is going to be so much harder. Those who wait might grow frustrated with their pups easily and think that it is the dog's fault they aren't learning fast enough.
Start to train them from the moment you move them in. Rather than associating the different tricks with just you, they will associate this new behavior with their home. It is a great fresh start for them, especially when adopted, so take this opportunity to create good habits right away.
Be consistent! You can't get mad at your two-year-old dog that it can't walk on a leash still if you never trained it to properly use a lead.
This is the same for those that wait too long to 'punish' their dog. If you come home to find the garbage has been torn through, they're going to be excited to see you because they probably forgot about this. You might get upset at them right away, and then they will think that you are simply upset to see them.
Though dog-shaming videos and memes are cute, dogs aren't reacting this way because they feel remorse over what they did. The reaction of the sad face and low head are only because they see that you are upset, and they're scared. No matter how guilty they might appear to be, they are really just making that expression because they are afraid of your mood (Manning, 2014).

Chapter 9

Your Step-by-Step Training Plan

Congratulations on making it to this part of the book! You are almost ready to bring your puppy home. The most important part is that you create a schedule. We have a how-to on the way that you can do this, along with a few other key principles to remember before we finish!

Get to Know Your Dog

As we already mentioned, all dogs are different. When you first bring your dog home, make sure that you are doing your best to really get to know them. They will have their own fears, strengths, weaknesses, and interests. Some dogs will be the type that always want to be outside, and others will be perfectly fine with sitting on the couch and sleeping all day. While we did our best to ensure this would be a comprehensive guide for all new pet owners to know how to train their dogs, you should still look into further research to discover the methods that you could use for specific breeds. Though it doesn't matter as much what random facts you might know about your dog, knowing their history, their intended breeding purpose, and so on can really help you better understand why your dog might operate in the way that it does.

Create a Schedule

It is important that you keep your dog on a schedule, just as you would a baby or a toddler. It is not only a good way for you to get into better habits, but it is the method that your dog is going to be able to learn how to function properly as well.

You should start to feed your puppy four times a day until they are around three months old. The best time to do it is when they first wake up and a couple of hours before bed. Then, you can do it twice in between whenever it is convenient for you, so long as it has been spaced out.

Next, until they are six months, drop down to three times a day. The same rules apply when you are figuring out when to feed them. It should be at times that are convenient for you in order to ensure that it is a schedule you can keep up with. For example, if you are usually heavy into work or studying around noon every day, don't try to disrupt your schedule with a feeding here. Though it might be a better time for your pup, if it is not a good time for you it might get forgotten or overlooked, resulting in inconsistencies which confuse your puppy.

Once you get them fixed, you can cut it down to twice a day. Make sure that you adhere to serving suggestions on the packaging of your selected food so as to not overfeed them. eight hours is usually a good time to feed them. After they have eaten, you should take them out around two hours later, though when training puppies, every hour is ideal if you can. They have smaller bladders and aren't going to be able to hold it. As they grow older and learn to control their bladder movements, you can reduce these time periods so that they learn how to resist the urge to go when they have to.

Here are all the things that you will want to include in your schedule for a puppy under six months:

- Three to four feedings

- Five potty breaks

- Training session

- Exercise routine

- Playtime

- Bed time

Again, it will really be dependent on your specific and individual schedule when these times might be. Stick to what works best for you and always consider your dog's needs. Here's an example of what a specific schedule might look like:

- 8 AM: potty break
- 8:30 AM: feed
- 10:30 AM: potty break
- 11 AM: training session
- 1 PM: feed
- 1:30 – 1:45: crate
- 2:00 PM: playtime
- 3 PM: potty break
- 3:30 PM – 3:45: crate
- 4 PM: potty break
- 5 PM: feed
- 7 PM: potty break + 30 minute walk
- 8 PM: feed
- 9 PM: playtime
- 10 PM: potty break
- 10:15 PM: bed time

This can seem like a lot, but feedings and potty breaks will be reduced as your puppy grows. Play around with your schedule at first, but after about a week of being in their new home, your puppy should have a set schedule that they can depend on.

Keep it Simple

Don't overly confuse your dogs. Always look for ways to keep things simple, short, and sweet because this is the method that they are going to be able to learn from the best. If you discover that they are having trouble learning a new trick, it could be that you are making it too complicated. If you find that it is too hard to keep them on a schedule, it might be something too complex for you. Though this book has been comprehensive and long, it is also not as hard as it might seem to keep your dog on a schedule and properly trained. With effort and some troubleshooting, you'll be able to find the perfect routine for you and your furry friend.

Work on Rewards

Remember that positive reinforcement is the center of proper training. When you are able to provide them with rewards they really like, you'll have exceptional results. Experiment with different treats to find the ones they like the most. The more they like a treat, the more likely they are to respond when you are training them! Rewards aren't just in the form of treats either. They're also the positive words and pets that you use for them. Notice the spots that they like scratched or pet the most. Do they prefer their belly rubbed? Their ears massaged? Their chins scratched? When you know your dog's favorite rewards, you can use these to get them to act the way that you want.

Repeat, Repeat, Repeat

Just because your dog successfully does a trick one time doesn't mean that they already know how to do this. They will still need repetition to ensure that they always really understand the trick you are trying to get them to do. though it can seem tedious and at times exhausting, go over and over and over the same trick to make sure that your dog has been able to properly learn it. Like we mentioned earlier, useless repetition will only set them back in their training, so make sure that you aren't repeating something that isn't working.

Check-In with Your Dog

Always consider your dog's needs first and foremost. Get to know their emotions. What do they do when they're anxious? How do they react when they're scared? Are they getting enough attention? Are you showing enough affection? Though it might seem easy to really understand a dog, they are more complex than you'd think! Just because something works for you doesn't mean it does for your dog. When things aren't going right, ensure that they are happy and taken care of. You are their caretaker, so you will always want to do regular check-ups with them to make sure you are providing the best care possible!

Conclusion

Don't feel nervous about getting a puppy, it should be exciting! At the same time, remember that it is crucial you train your puppy in a responsible way and never overlook the important skills they need to be a healthy dog.

Start by bringing home a dog that will work for you in the first place. Though you might like the look of a certain breed, an Australian Shepherd isn't good for someone who doesn't go outside often and enjoys sitting on the couch. Vice versa, if you are someone who enjoys an active lifestyle activities such as exercising, hiking, and so on, a mini poodle that sleeps for 20 hours a day isn't the top choice.

After you've picked your dog, it is time to get your home ready. Pick everything up that could be dangerous, and provide them with a safe space that they don't have to feel uncomfortable in. Give them necessary items and pick out the toys that will help them grow and learn. Give them plenty of treats and water, as well as carefully selected food.

Ensure that you start training from the moment that you get home. Teach them how to sit, and how to go to the bathroom outside right away. Every other trick that you teach them will likely build from sitting down first, and this is a great way that you can make them calm and collected whenever you need them to act that way.

Socialize them from an early age as well. They'll need to get used to controlling their excitement in public, and it is essential that we teach them good behavior with other dogs and people from the very start. This is especially important for aggressive dogs. You have to also ensure you are getting them out of the house to tire them out. Active and ornery dogs may go a bit stir-crazy when you don't get them outside and playing around enough.

Once you have a healthy routine and can start to socialize them properly, it is time to add in more challenging skills. These are things like rolling over or laying down. Once they are able to learn one trick, it will be easier for them to learn even more.

After all of this, ensure that you are giving them the right amount of exercise. Though you might be playing with them and training them at the same time, they will need separate moments of exercise. This could be in the form of agility training or going for basic walks.

At the same time, you'll need to ensure you take them to the vet for regular check-ups and that all of their health needs are being met. Give them proper baths and grooming and don't forget about their need for flea or tick treatments as well.

Never forget the common mistakes that other dog owners might make. Each dog is unique and needs to be treated this way, so even though something might work for one dog, that doesn't mean it will for another.

Your dog is going to be your best friend. They will think about you all day long when you are not home. You are the one that they care about more than anything and they put all of their trust in you. Give them the best chance possible at a happy and healthy life by always properly training them.

References

ASPCA. (n.d.). Pet Statistics. Retrieved from https://www.aspca.org/animal-homelessness/shelter-intake-and-surrender/pet-statistics

ASPCA. (n.d.) Vaccinations for Your Pet. Retrieved from https://www.aspca.org/pet-care/general-pet-care/vaccinations-your-pet

Becker, K. (2012). Want a Well-Behaved Dog? Do More of This and Less of That. Retrieved from https://healthypets.mercola.com/sites/healthypets/archive/2012/08/03/positive-reinforcement-dog-training.aspx

Bovsun, M. (2019). How to Potty Train a Puppy: A Comprehensive Guide for Success. Retrieved from https://www.akc.org/expert-advice/training/how-to-potty-train-a-puppy/

Dayton, R. (2016). New Study Shows Potential Benefits Of Spaying/Neutering Dogs After Age 1. Retrieved from https://pittsburgh.cbslocal.com/2016/11/14/new-study-shows-potential-benefits-of-spayingneutering-dogs-after-age-1/

Finlay, K. (2017). Can Dogs Taste? Retrieved from https://www.akc.org/expert-advice/lifestyle/can-dogs-taste/

Greenwood, A. (2016). Here Are The Surprising, Preventable Reasons Why People Give Up Their Pets. Retrieved from https://www.huffpost.com/entry/aspca-report-pets-given-up-to-shelters_n_56896846e4b06fa68882a134

Keliher, I. (n.d.). How to Choose the Perfect Dog Name (with Science). Retrieved from https://www.rover.com/blog/dog-name-advice/

Manning, S. (2014). Behaviorists: Dogs feel no shame despite the look. Retrieved from https://www.usatoday.com/story/news/nation/2014/02/26/dogs-shame-guilty-look/5833395/

The Humane Society. (n.d.) Crate training 101. Retrieved from https://www.humanesociety.org/resources/crate-training-101

The Humane Society. (n.d.) Fact Sheet: Puppy Mills and Pet Stores. Retrieved from https://www.humanesociety.org/sites/default/files/docs/pet-stores-puppy-mills-factsheet.pdf

The Humane Society. (n.d.). Pets by the numbers. Retrieved from https://www.animalsheltering.org/page/pets-by-the-numbers

Walsh, K. (2017). Possible Health Isuses in Common Dog Breeds. Retrieved from https://www.healthline.com/health/dog-breeds-and-health-issues